Battlefield Weapons Systems & Technology Volume III

AMMUNITION

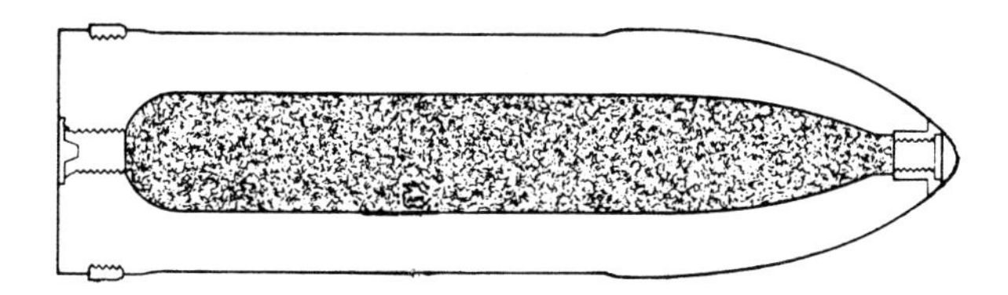

(including grenades and mines)

Other Titles in the Battlefield Weapons Systems and Technology Series

General Editor: Colonel R G Lee OBE, Military Director of Studies at the Royal Military College of Science, Shrivenham, UK

This new series of course manuals is written by senior lecturing staff at RMCS, Shrivenham, one of the world's foremost institutions for military science and its application. It provides a clear and concise survey of the complex systems spectrum of modern ground warfare for officers-in-training and volunteer reserves throughout the English-speaking world.

Volume I — Vehicles and Bridging—I F B Tytler *et al*

Volume II — Guns, Mortars and Rockets—J W Ryan

Volume III — Ammunition, including Grenades and Mines—K J W Goad and D H J Halsey

Volume IV — Nuclear, Biological and Chemical Warfare—L W McNaught

Volume V — Small Arms and Cannons—C J Marchant Smith and P R Haslam

Volume VI — Command, Communication and Control Systems and Electronic Warfare—A M Willcox, M G Slade and P A Ramsdale

Volume VII — Surveillance and Target Acquisition Systems—W Roper *et al*

Volume VIII — Guided Weapons—D E Tucker and T K Garland-Collins

Volume IX — Military Data-processing and Microcomputers—J W D Ward and G N Turner

Volume X — Basic Military Ballistics—C L Farrar and D W Leeming

For full details of these and future titles in the series, please contact your local Brassey's/Pergamon office

Related Titles of Interest

Garnell — Guided Weapons Control Systems (2nd ed.)

Hemsley — Soviet Troop Control. The Role of Automation in Military Command

Lee — Introduction to Battlefield Weapons Systems and Technology

Morris — Introduction to Communication, Command and Control Systems

AMMUNITION

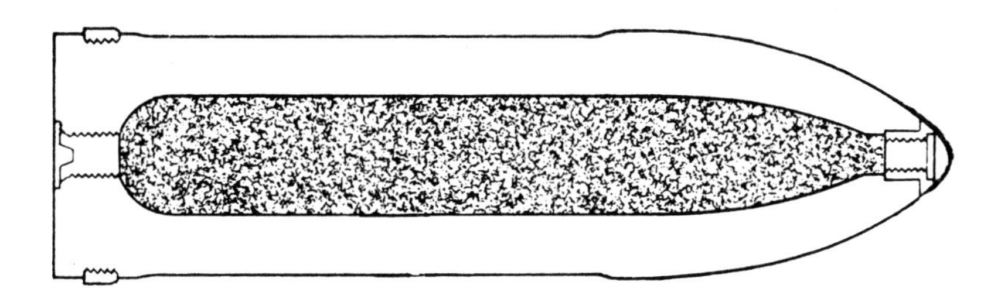

(including grenades and mines)

K J W GOAD and D H J HALSEY
Royal Military College of Science, Shrivenham, UK

BRASSEY'S PUBLISHERS LIMITED
a member of the Pergamon Group

OXFORD · NEW YORK · TORONTO
SYDNEY · PARIS · FRANKFURT

U.K.	BRASSEY'S PUBLISHERS LIMITED, a member of the Pergamon Group, Headington Hill Hall, Oxford OX3 0BW, England
U.S.A.	Pergamon Press Inc., Maxwell House, Fairview Park, Elmsford, New York 10523, U.S.A.
CANADA	Pergamon Press Canada Ltd., Suite 104, 150 Consumers Road, Willowdale, Ontario M2J 1P9, Canada
AUSTRALIA	Pergamon Press (Aust.) Pty. Ltd., P.O. Box 544, Potts Point, N.S.W. 2011, Australia
FRANCE	Pergamon Press SARL, 24 rue des Ecoles, 75240 Paris, Cedex 05, France
FEDERAL REPUBLIC OF GERMANY	Pergamon Press GmbH, 6242 Kronberg-Taunus, Hammerweg 6, Federal Republic of Germany

First edition 1982

Library of Congress Cataloging in Publication Data

Goad, K. J. W.
Ammunition (including grenades and mines)
(Battlefield weapons systems & technology;
v. 3)
Includes index.
1. Ammunition—Handbooks, manuals, etc
I. Halsey, D. H. J. II. Title. III. Series.
UF700.G6 1982 623.4 81-23411

AACR2

British Library Cataloguing in Publication Data

Goad, K. J. W.
Ammunition (including grenades & mines).—
(Battlefield weapons systems & technology; v.3)
1. Ammunition
I. Title II. Halsey, D. H. J. III. Series
623.4'55 UF700

ISBN 0-08-028326-8 (Hard cover)
ISBN 0-08-028327-6 (Flexi cover)

In order to make this volume available as economically and as rapidly as possible the authors' typescripts have been reproduced in their original forms. This method unfortunately has its typographical limitations but it is hoped that they in no way distract the reader.

The views expressed in the book are those of the authors and not necessarily those of the Ministry of Defence of the United Kingdom.

Printed in Great Britain by A. Wheaton & Co. Ltd., Exeter

Foreword

There are many publications dealing with ammunition available on the market today, but most adopt a catalogue format and are restricted to dealing with one type or nature of ammunition such as small arms ammunition, or artillery ammunition, or mines and so on. They are thus most useful to the collector or enthusiast whose hobby and main interest is in a particular aspect of ammunition.

This Volume concentrates on the requirements, methods of operation and design principles of the range of different kinds of ammunition, making reference to specific natures of ammunition or equipments for the sake of illustration only. As such it is intended primarily to assist the young and middle piece army officer to widen his knowledge of a vital and fundamental commodity of his everyday military life.

Preface

This series of books is written for those who wish to improve their knowledge of military weapons and equipment. It is equally relevant to professional soldiers, those involved in developing or producing military weapons or indeed anyone interested in the art of modern warfare.

All the texts are written in a way which assumes no mathematical knowledge and no more technical depth than would be gleaned from school days. It is intended that the books should be of particular interest to army officers who are studying for promotion examinations, furthering their knowledge at specialist arms schools or attending command and staff schools.

The authors of the books are all members of the staff of the Royal Military College of Science, Shrivenham, which is comprised of a unique blend of academic and military experts. They are not only leaders in the technology of their subjects, but are aware of what the military practitioner needs to know. It is difficult to imagine any group of persons more fitted to write about the application of technology to the battlefield.

This Volume

There are many publications dealing with ammunition available on the market today, but most adopt a catalogue format and are restricted to dealing with one type or nature of ammunition such as small arms ammunition, or artillery ammunition, or mines and so on. They are thus most useful to the collector or enthusiast whose hobby and main interest is in a particular aspect of ammunition.

This Volume on Ammunition concentrates on the requirements, methods of operation and design principles of the range of different kinds of ammunition, making reference to specific natures of ammunition or equipments for the sake of illustration only. As such it is intended primarily to assist the young and middle piece army officer to widen his knowledge of a vital and fundamental commodity of his everyday military life.

Shrivenham. November 1981 Geoffrey Lee

Acknowledgements

The authors are grateful for the valuable assistance given to them by a number of Ministry of Defence (Army Department) Branches; in particular the Directorate of Land Service Ammunition.

They wish also to acknowledge the expertise and work of the RAOC Officers (Col (Ret'd) C J Isaac, Lt Col (Ret'd) P G Clayton, Lt Col (Ret'd) M Haywood and Lt Col J F F Sharland) who held the appointment of "DS Ammo" at The Royal Military College of Science during the intervening years between Col Halsey's relinquishment of the post in 1968 and Lt Col Goad's assumption of it in 1980.

Finally, they wish to thank Mrs Linda Hammond for typing the book.

Shrivenham
September 1981

KJWG & DHJH

Contents

List of Illustrations

Chapter 9

Chapter 10

Chapter 11

Chapter 12

Chapter 13

Chapter 14

Chapter 15

Chapter 16

Chapter 17

1

Introduction to Ammunition

INTRODUCTION

In its widest sense both literally and metaphorically the word ammunition covers anything that can be used in fighting. The word is thought to have originated from the Latin words "moenia" meaning walls and "munire" meaning to fortify, although the actual word ammunition as used today is probably an English corruption of the obsolete French words "l'a munition" or "la munition". The French used l'a munition to embrace all the material used for war, although in time it was used more specifically to cover powder and shot.

For military purposes, ammunition can be defined as any munition of war whether defensive or offensive, or any component whether filled or intended to be filled with explosive, smoke, chemical, incendiary, pyrotechnic or any other substance designed to affect the target, including any inert or otherwise innocuous training, practice or drill replicas and expedients. In other words, ammunition is the generic term for all devices from a pistol bullet to a high velocity kinetic energy anti-tank projectile, from a hand thrown grenade with a range of 25 metres to a heavy artillery shell with a range of 30 kilometres and from a simple illuminating rocket to an inter continental ballistic missile. Figure 1 shows some of the range of items that are classed as ammunition.

The purpose of ammunition is to provide a required effect at a selected target. The means of delivering ammunition to the target are many and varied. They range from explosive projection from a gun or mortar, to self propulsion as in guided and unguided rockets and torpedoes, to being thrown either by hand or by non explosive mechanical means. Ammunition can also be transported to the target area and there released as in an aircraft bomb, or it can be hand placed like a demolition charge or, yet again, it can be left in a selected area to be initiated remotely or by the target itself as in the case of a mine or booby trap. There are many other ways of delivering ammunition to a target, but the more conventional delivery means most used by land forces, guns and mortars, are described in detail in companion volumes to this book. It is important, however, to appreciate from the outset that ammunition is but part of a system and should not be considered in isolation from the other components that make up the complete weapon system. The ammunition designer must pay due regard to the means of delivery.

The most effective way of disabling a target, whether animate or inanimate, is to apply energy to it in some way and as rapidly as possible. The best conventional way to do this is to use explosives and within the definition given above, all service stores containing explosives are classed as ammunition.

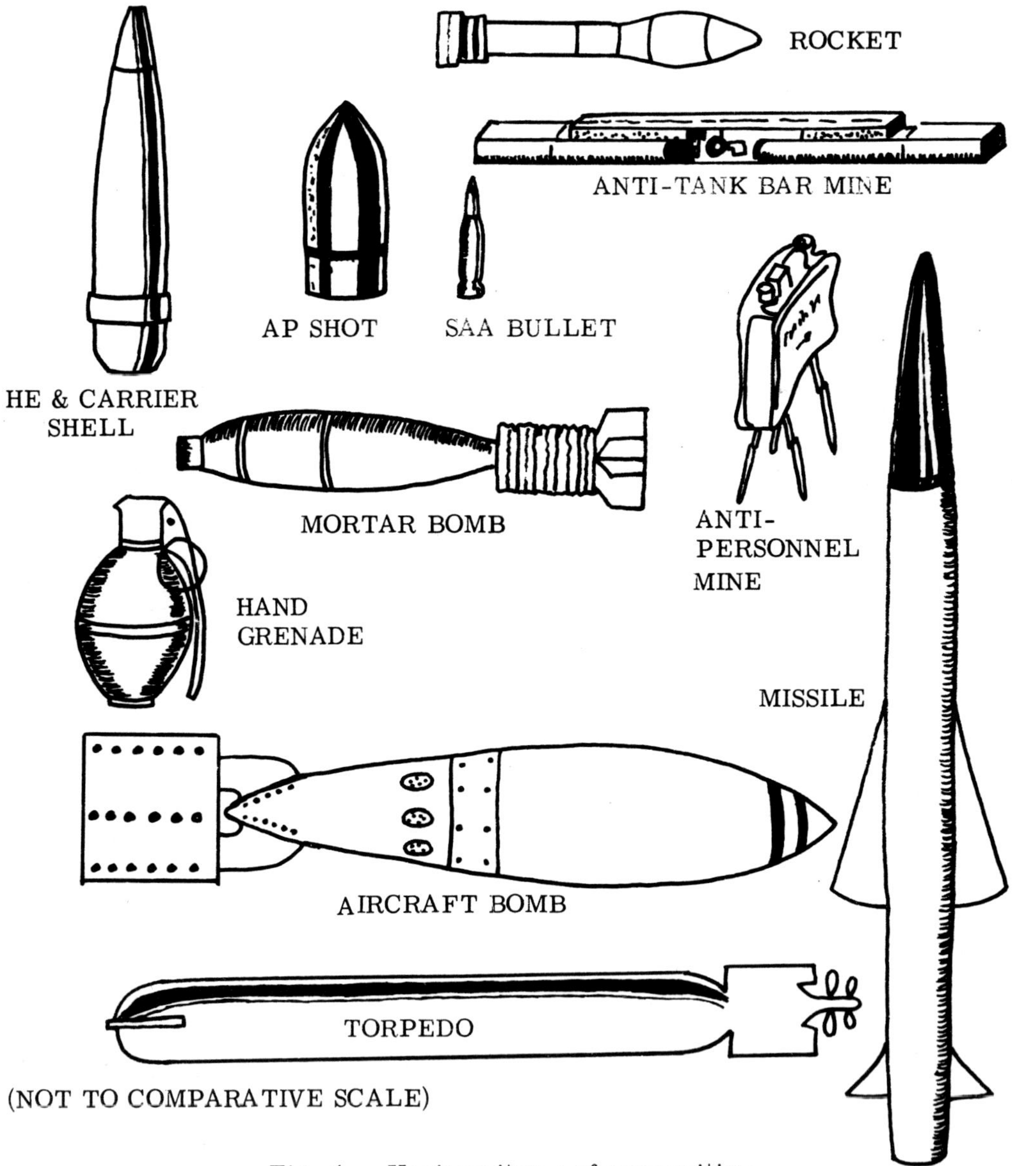

Fig. 1. Various items of ammunition

EXPLOSIVES

Explosives rank as one of the greatest discoveries which have exerted a profound influence on the character of warfare. Explosives can be used to project a solid shot at high velocity at a target relying on the kinetic energy (KE) that has been so imparted to it to achieve the target effect. This explosively propelled kinetic energy projectile dissipates its energy at the target on impact and does not require a triggering device to control the energy release. Alternatively, chemical energy (CE) can be used in the form of an explosive substance which has considerable potential energy relative to its mass which can be released when suitably initiated. The energy release trigger mechanism is known more familiarly as a fuze. Explosives, therefore, are a convenient source of energy for use in ammunition, both to achieve an effect against the target, as well as propelling it to the target where this is necessary. Different types of explosives, however, are required to perform these two functions; the characteristics of "high" explosives being more generally suited to achieving target effects, whilst "low" explosives are used more generally as propellant charges. The different characteristics of each are explained in detail in Chapter 4. Nevertheless, it is appropriate here to offer a definition of an explosive and highlight the differences between "high" and "low" explosives.

Explosion and Detonation

An explosive has already been described as a substance with considerable potential energy relative to its mass which can be released when suitably initiated. On being initiated it exerts a sudden and intense pressure on its surroundings, the pressure being developed by the decomposition of the explosive into gas with a simultaneous liberation of heat. The difference between low and high explosives is the rate at which they decompose when initiated. Whilst a low explosive burns (deflagrates), albeit extremely rapidly, a high explosive "detonates".

Detonation is a word that derives from the Latin "de" (down) and "tonare" (to thunder), and detonating high explosives achieve their disruptive effect by a shattering process. Detonation takes place within the explosive and arises from an instantaneous rearrangement of the atoms caused by an intense shock wave, itself generated and maintained by the detonation phenomenon.

Explosive Trains

An ammunition contains an explosive train which is either an igniferous or a disruptive train; a "complete round" of ammunition which has to be fired from a gun, contains both.

An igniferous or burning train uses low explosives and is used for propulsion. In this type of train the explosives achieve their propelling effect by a burning process. Burning takes place on the surface of the explosive. The rate of the event is essentially controlled by the available surface area of the charge weight being used, although it is influenced by other factors such as confinement. By confining the explosive, pressure is increased which in turn increases the rate of burning.

A disruptive or detonating train is used to achieve target effects, and in munitions such as mines and grenades which do not have to be explosively propelled to the target, it is the only explosive train used. In this type of train the explosives achieve their effect by a detonating, or shattering, process as explained above. The rate of this event is essentially an inherent property of the explosive being used - a property known as "velocity of detonation" - although again it is obviously influenced by other factors.

Elements of an Explosive Train

In any explosive train, be it igniferous or disruptive, it is usually possible to identify three separate elements: an initiator, an intermediary and a main filling. Typical components found in these two types of train, together with an indication of their relative sensitivity (described more fully below under properties of explosives) are shown in Fig. 2.

(a) Igniferous Train (BURNING) for Propulsion

Electrical or Mechanical Impulse (Striker))	)	
Cap Composition containing a very)	- Initiator)	- Primer
sensitive high explosive (10))	)	
Gunpowder (60)	- Intermediary)	
Propellant (100)	- Main Filling	

(b) Disruptive Train (DETONATING) for Target Effects

Electrical or Mechanical Impulse (Striker),)	
or some form of igniferous delay)	- Initiator
Detonator (10))	
CE (Composition Exploding: Tetryl) (60)	- Intermediary
High Explosive (TNT: RDX/TNT: HMX etc.) (100)	- Main Filling

Fig. 2. Elements of igniferous and detonating trains

Properties of Explosives

The important properties of explosives with which the ammunition designer is concerned are sensitivity, power, velocity of detonation, brisance and compatability.

Sensitivity gives some measure of the care required, and safeness, in handling the explosive. It is quoted as a "Figure of Insensitivity" (F of I) quantitatively related to the sensitivity of Picrite which has a nominal F of I of 100. Initiators have a low F of I, whilst main fillings are relatively insensitive with Fs of I typically of the order of 100 or more (see Fig. 2 above).

Power gives a measure of the energy available from an explosive and is based on the volume of gas and quantity of heat released per unit weight of the explosive. Quantitatively it is again related to Picrite, which has a nominal power of 100.

In general, main fillings of high power are used, but with initiators and intermediaries other properties, such as sensitivity, are of more importance than power. Power is not in this context used in the mechanical sense, and it is not a rate of doing work.

The velocity of detonation (V of D) of an explosive, expressed rather simplistically, is an indication of the rate at which it releases its energy. High explosives of low V of D are used as lifting charges - for cratering - where a low energy release rate is required, whilst a high V of D would be used for an intense shattering effect or where other effects are required (high fragment velocities from shells, high collapse rates with shaped charge liners - see Chapters 3 and 8 - and other such factors). Vs of D are measured absolutely in metres per second and typically range from 2000 to over 9000 metres per second.

Brisance is a very loose qualitative property which gives a somewhat abstract indication of the "snap" of the explosive. It can be regarded as the product of the power and the velocity of detonation of the particular explosive.

Finally, it is vital that explosives are chemically compatible with all materials with which they might come into contact in any piece of ammunition. Long term compatibility of explosives and certain materials (notably metals and plastics) over a wide range of climatic conditions, is probably one of the most constant and difficult problems facing the designer.

GUN LOADING SYSTEMS

Certain types ("natures") of ammunition have to be propelled to their target from a projector; be it a mortar, a soldier or a gun. Leaving aside human projection which does not, by and large, impose many constraints on the ammunition designer, projection from a gun or mortar does pose problems. One of these problems is to ensure that the gases evolved from the propellant charge are so sealed in the gun or mortar that they cannot escape, and can thus only act against the projectile and in doing so propel it towards the target. This sealing of the weapon is known as "obturation". Obturation can be defined as the prevention of the uncontrolled escape of propellant gases: with conventional tube launched (from guns and mortars) projectiles this usually implies the prevention of the uncontrolled escape of propellant gases to the rear.

In mortars which are muzzle loaded, the rearward escape of gases is prevented by a closed breech end. In this case the obturation problem is to prevent the uncontrolled escape of propellant gases forward past the bomb. For conventional breech loaded (BL) guns, the problem is how to seal the chamber of the gun effectively when the breech has to be rapidly and frequently opened to insert another round. The methods of achieving obturation are discussed in some detail in Chapters 5 and 12; suffice it to say at this stage that the method of obturation determines the structure of the ammunition.

Cased Charge Obturating System

If obturation is a function of the ammunition, the propellant charge must be enclosed in a metal case. Where obturation is done by the ammunition, the ammunition is known as Quick Firing (QF) ammunition; presumably because only one piece of ammunition, the round and the propellant charge enclosed in an attached metal case, had to be loaded, and this could be done quickly. Obturation is obtained in the following way. When the charge starts to burn the propellant gases try to escape both forwards past the loosely positioned projectile, and rearwards through the gap between the outside of the case and the inside of the gun chamber wall. As the chamber pressure increases the projectile is pushed forward and the driving band fully engages the rifling in the barrel: this provides a forward gas seal which is maintained throughout shot travel. At the same time the case expands radially, first elastically and then plastically, until it closes completely with the chamber wall. Gas flow, which may have occurred down a short length of the outside of the case, is then checked and complete obturation is achieved. As the shot leaves the muzzle of the gun, the chamber pressure falls and the case contracts elastically to allow extraction and subsequent ejection. (For diagram see Chapter 5, Fig. 2).

Bagged Charge Obturating System

If on the other hand obturation is a function of the gun, the propellant charge does not have to be enclosed in a metal case. In this instance obturation is achieved by the following process. When the main charge starts to burn the propellant gases try to escape, again both forwards past the loosely positioned projectile and rearwards past the loosely fitting obturator pad assembly. As the chamber pressure increases the projectile is pushed forward and the driving band fully engages the rifling in the barrel: this provides a forward gas seal which is maintained throughout shot travel. At the same time pressure is exerted on the steel mushroom head in front of the obturator pad: under this pressure the mushroom head squeezes the obturator pad, which distorts radially outwards to provide a seal against its steel seating, and complete obturation is achieved. At shot ejection the chamber pressure falls and the obturator pad releases its stored energy to revert to its original shape. See Chapter 5, Fig. 1. Bagged charge obturation is often referred to as a Breech Loading (BL) System. Figure 3 shows the various types of QF and BL ammunition. In fact the terms QF and BL no longer have a place in describing modern ammunition; the terms can be positively misleading. All QF systems are breech loading, and breech loading was introduced to give a higher rate of fire (higher than rifled muzzle loading!). (For diagram see Chapter 5, Fig. 1).

HISTORICAL DEVELOPMENT

Early Ammunition and Weapons

To appreciate fully the high standards of safety and performance of modern ammunition, it is necessary to look, albeit very briefly and very much in outline, at the evolution of ammunition and weapons to highlight some of the more significant developments.

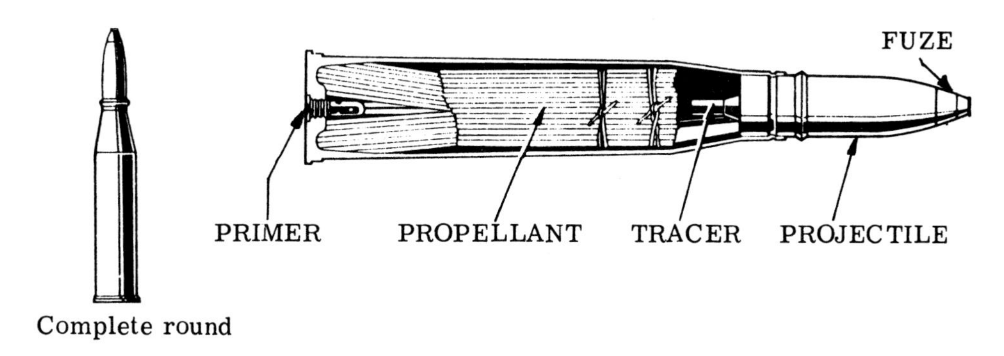

QF FIXED AMMUNITION

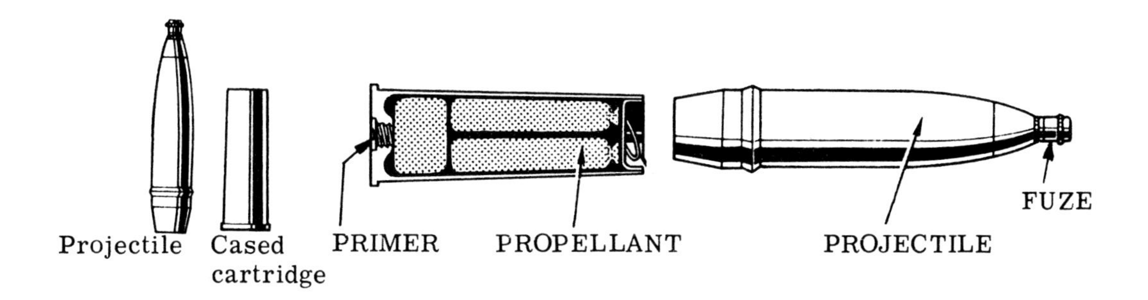

QF SEPARATE LOADING AMMUNITION

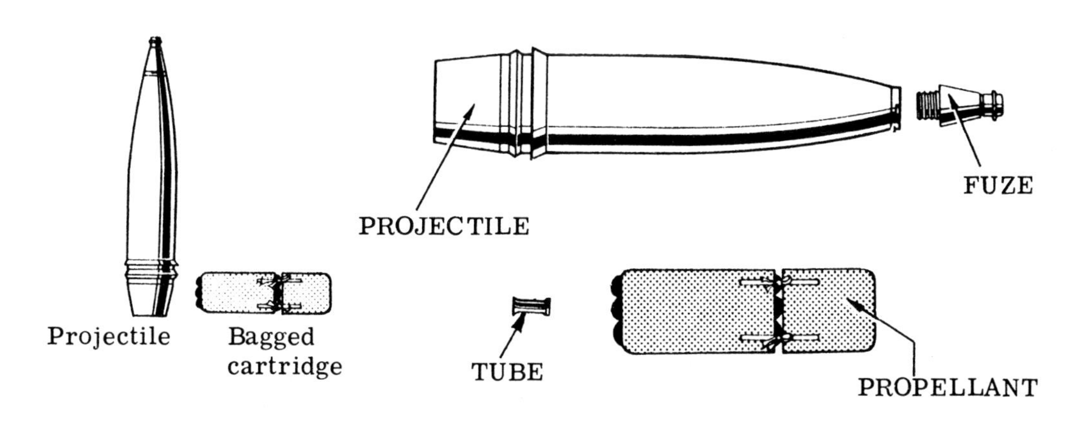

BL AMMUNITION

Fig. 3. Various types of gun-fired ammunition

The requirement for ammunition undoubtedly arose from Man's need to kill, either to feed himself or to protect himself. To overcome the problem of distance between himself and his quarry, or to put distance between himself and his adversary, he threw simple missiles like stones and spears. In time, he moved onto bows and slings which not only further increased distance ("range") but also added force ("hitting" or "stopping" power) to his missiles. Even so, these weapons and missiles were ultimately dependent on the strength of each individual man. There followed, therefore, the development of simple mechanical devices to store energy which, when released, threw larger and heavier missiles with greater force over longer distances than man could do using his own strength. The crossbow, the catapulta and the ballista (shown in Fig. 4) are examples of early mechanical delivery means.

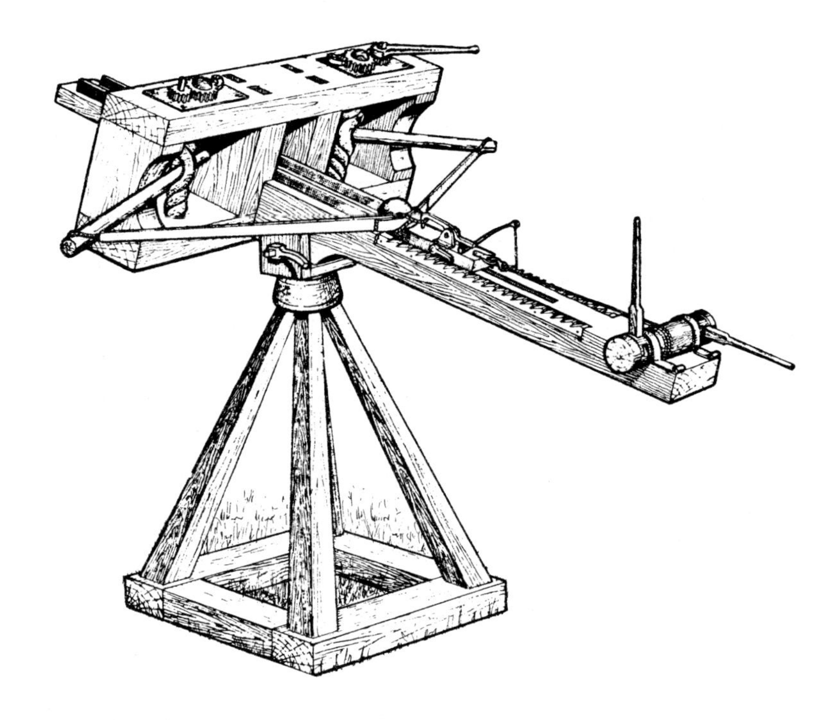

Fig. 4. The Ballista

Gunpowder

The most significant developments in both ammunition and weapons followed the refinement of gunpowder in the thirteenth century, which enabled it to be used in time as a propellant explosive to propel missiles and other projectiles over long distances. Gunpowder revolutionised the concepts of warfare and changed the strategies and tactics of mediaeval Europe as fundamentally as the advent of the nuclear age changed the twentieth century world after the first atom bomb was dropped on Japan in 1945.

The industrial revolutions in Europe in the eighteenth and nineteenth centuries which gave rise to the chemical industry, also provided the necessary impetus for the development of explosives for industrial use. Inevitably, their potential for military use was quickly realised and exploited. The manufacture of high explosives, which detonated rather than exploded, was a particularly significant development of this period, and marked the beginning of conventional modern warfare.

Cannons and Guns

The potential of gunpowder to propel a heavy missile some distance led to the development of the cannon and the other great guns of artillery, as well as to small arms.

Cannons and other large guns of artillery were also known, and still are, as pieces of "ordnance", a reference to the Board of Ordnance which under the Master General of Ordnance controlled all artillery weapons in the British Army until after the Battle of Waterloo in 1815.

The earliest cannons were made of longitudinal bars hooped around with wrought-iron coils rather like wooden staves are hooped around to form a wooden barrel of the type to hold water or beer. This, incidentally, is probably how the "gun barrel" came to be so called. The early cannons had smooth bore barrels into which the gunpowder propelling charge was inserted through the muzzle, followed by a wad or wooden tampion, and rammed home. The projectile used at this time was a solid iron ball, also loaded in through the muzzle and then rammed against the propelling charge. A small quantity of gunpowder was poured into a touch-hole at the rear of the cannon above the gunpowder charge and fired by means of a slow-match. From this basic design, the cannon gradually evolved from a smooth bore muzzle loading weapon to the rifled barrel breech loading artillery guns of today. Similarly, small arms evolved from smooth bore muzzle loading to rifled barrel breech loading weapons. Mortars, however, remain muzzle loading and generally smooth bore. Indeed many countries (but not the United Kingdom) have smooth bore barrels for their tank guns, but these are breech loading.

The development of guns and small arms is covered in much greater detail in Volume II - Guns, Mortars and Rockets and Volume V - Small Arms and Cannons in this Battlefield Weapons Systems and Technology series.

Early Projectiles

Ammunition evolved alongside gun developments; an advance or innovation in one prompting a further development in the other and vice versa.

The early arrow-shaped projectiles were replaced by large stones and round solid iron shot which were found to be more effective at demolishing fortifications which were the most difficult targets to defeat at that time. (It is a historical irony, that modern ammunition designers have returned to the arrow-shaped projectile to defeat today's most difficult target in the land battle, the tank; see Chapter 8).

Attempts were made to fill iron shot with gunpowder and to provide it with an elementary fuze, the action of firing was to ignite a piece of slow-match inserted in the side of the shot. The hazards involved were great and the variations in timing the burst were highly unpredictable. "Prematures", whereby the shot exploded whilst still in the barrel were a common occurrence. Solid iron shot, therefore, continued to be used to breach defence works and fortifications, whilst "langridge" (or "langrage") was used against soldiers in the open. Langridge was a shot containing irregular pieces of iron and metal such as bolts, nails and other pieces of scrap.

Fuzes

In the eighteenth century a time fuze of sorts was introduced in an attempt to burst the shot at the optimum time. This fuze was essentially a wooden peg which was grooved spirally around its exterior. The grooves were lined with a core of gunpowder and could be cut at the required groove, thus controlling, very roughly, the time of burning. From this unsophisticated early attempt to control the time and place of projectile bursting, evolved the complex, intricate and reliable fuzes of today: see Chapter 11. This is also a convenient place to stress the correct spelling of the word fuze when it is used in the context of a mechanism to cause a projectile to function at a required time and or place. For this definition, it is spelt with a 'z'. When spelt with an 's' (fuse) it has different meanings; as a verb transitive it means to melt: to liquefy by heat: to cause to fail by melting of a fuse as in electricity. Used as a noun, fuse is a train of combustible material in a waterproof covering, used with a detonator to initiate an explosion; for example, safety fuse.

Modern Projectiles

In 1784, Lieutenant Henry Shrapnel invented a hollow round shot containing a fuze, a powder bursting charge and a quantity of small shot pellets or "shrapnel" as they became known as. The shrapnel shell was designed to burst in the air above enemy positions, and the shot pellets which it contained, rained down on the enemy. This projectile is significant because it enabled enemy troops to be engaged at ranges greatly in excess of those attained by other anti-personnel projectiles, such as langridge, cannister and grape-shot that were in use at that time. These anti-personnel projectiles started to spread their contents out immediately on leaving the muzzle of the gun, like the pellets of a modern hand held shot gun.

The introduction of rifled barrels which spun projectiles, together with fuze and fragmentation improvements led to the development of cylindrical projectiles which replaced the spherical shot. The cylindrical shell, known as the "common shell" was in essence the same basic design of projectile as today's HE shell, as can be seen in Fig. 5. The common shell had a shaped, tapered head (the "ogive"), a driving band and a fuze; all these remain features of the modern HE shell.

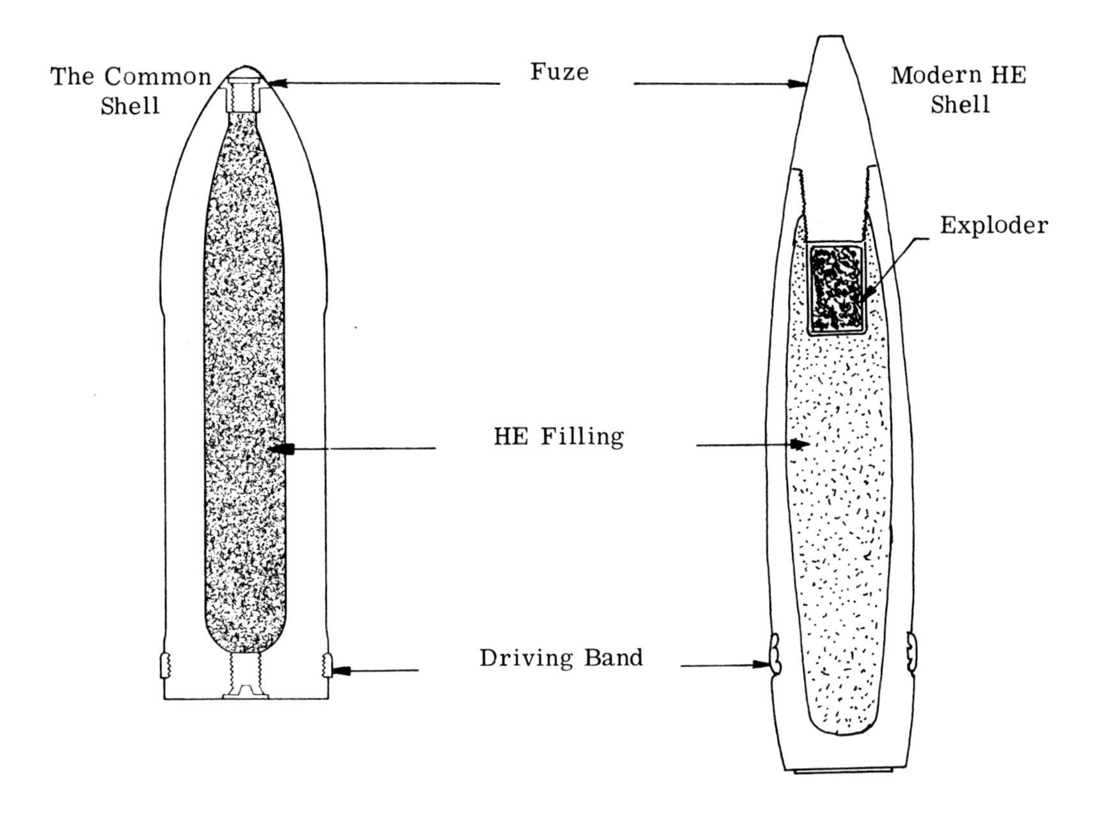

Fig. 5. The "common shell" and a typical modern HE shell

SUMMARY

Ammunition is a very wide subject and is intimately and inseparably bound up with the whole range of weapons that have been developed and used throughout history.

The word also covers all stores employed offensively or defensively in the prosecution of land warfare, but which are not necessarily associated with any particular armament.

Explosives are a convenient source of energy and are widely used in ammunition, both for propulsion and for achieving a disruptive effect against a target. The progress and evolution of science, particularly during the eighteenth and nineteenth centuries as a consequence of the advances made during the industrial revolution in Europe, led to the introduction of the first truly modern weapons and ammunition during the latter half of the nineteenth century.

The evolution of the fuze, the adoption of rifled, breech loading guns in preference to smooth bore muzzle loading weapons, the supersession of the only known explosive, gunpowder, by more efficient low explosives and more powerful high explosives, all combined to affect profoundly the conventional and established approach to the science of war.

These developments remain the bases on which modern ammunition and weapon technology rests.

SELF TEST QUESTIONS

QUESTION 1 Define the word ammunition.

Answer ....

....

....

....

QUESTION 2 What, in general terms, is the most effective conventional means of disabling a target?

Answer ....

....

....

....

QUESTION 3 Define an explosive.

Answer ....

....

....

....

QUESTION 4 Name the types of explosive, and the uses to which each type can be put.

Answer ....

....

....

....

QUESTION 5 Define an explosion.

Answer ....

....

QUESTION 6 Define a "high" explosive.

Answer ....

....

....

....

QUESTION 7 Name the types of explosive train.

Answer ....

....

....

....

QUESTION 8 What is meant by the term "brisance"?

Answer ....

....

....

....

QUESTION 9 Define "obturation".

Answer ....

....

....

....

QUESTION 10 Why are artillery guns sometimes referred to as "ordnance"?

Answer ....

....

....

....

ANSWERS ON PAGE 247

2

The Application of Fire to Targets

INTRODUCTION

The effectiveness of a complete weapon system can be measured and compared with other systems only if a framework is established in which certain basic characteristics are considered individually and together. The framework most frequently used by a weapon system designer on which to base an assessment of the effectiveness of his system is the concept of the overall kill chance. Within this concept, the chance of a kill, P_K, in the simple case of a one round engagement, can be expressed as a simple probability product rule:

$$P_K = P_H \times P_R \times P_L$$

Where P_H is the chance of a hit, P_R is the reliability of the weapon system, and P_L is the lethality of the warhead or ammunition. Cumulative probability rules can be applied for multiple round engagements.

A weapon system designer cannot meaningfully define a target by reference solely to its inherent characteristics. He must also consider both the scale and mode of attack and the probable circumstances of the engagement (e.g. is it a direct or indirect engagement?; what is the range?: and other such factors). For the ammunition designer, however, the first essential consideration is whether or not, given a random hit on the target, the terminal performance of the ammunition will produce the required effect. He is primarily concerned, therefore, with the lethality term, P_L, of the overall kill chance expression; reliability and the chance of a hit are more the concern of the designer of the complete weapon system. Lethality, as far as the ammunition designer is concerned, can be defined as the chance of a kill given a random hit on the target, and is simply the ratio of the vulnerable area of the target to the total presented area for a given power or weight of attack.

DAMAGE LEVELS

The ammunition designer must know the level of damage which he has to achieve against his target, since the lethality of the ammunition he is designing must be related to this. Three damage levels are usually considered: neutralisation, disablement and destruction. Neutralisation is a somewhat abstract damage level consideration since its effect stems from the apprehension induced in the target rather than to the casualty causing potential of the ammunition being fired at it. Hence it is people rather than material which can be neutralised, and this is impossible to quantify at all meaningfully. Neutralisation is not a damage level criterion against which the ammunition designer can design. Of the other two criteria, disablement and destruction, disablement of the target to prevent it from doing its job is the damage level the designer aims to achieve. The destruction of land force targets is seldom cost-effective, since it leads to substantial over-kill.

TYPES OF TARGETS

To establish the level of lethality the ammunition designer has to achieve, the full range of possible targets has to be identified and then examined for their vulnerability to different modes of attack.

The main types of targets with which land forces are confronted are people, armoured vehicles, structures, equipment and aircraft. This ambit of targets is not exhaustive, nor are the targets listed in any form of priority. It is, however, worth looking at them in some detail to identify and consider the factors which affect their vulnerability.

People

People appear as targets with varying degrees of attached protection. The type of clothing, and hence the protection it will afford them, will vary for many and diverse reasons; the theatre of operations, the season of the year, the type of work being undertaken and so on. Under most conditions men will be wearing steel helmets, although when employed in certain supply roles this may not always be true. Personal equipment will vary and this will also affect to an extent their vulnerable presented area: they may even be wearing bullet-proof vests or other special protective clothing.

People also appear as targets with varying degrees of "tactical" protection. They may be fully exposed, behind light cover, behind heavy cover, well dug-in and protected by fortifications or in buildings and other locally available protective hide-outs. Even when appearing fully exposed on open ground the degree of protection afforded to them will vary with the type of ground: lightly wooded, broken scrubland will offer more protection than flat open desert.

In addition to their worn protection and tactical protection, people appear in different attitudes: standing, kneeling or prone. They appear also in differing numbers: singly, tightly knit bunched groups or groups well spread out over a large area.

People, therefore, represent a complex target pattern, appearing in varying densities and, due either to their attached protection, tactical protection or attitude, with varying effective exposed areas.

Armoured Vehicles

Armoured vehicles are arguably the main type of targets that will have to be engaged on the modern battlefield. Their characteristics of mobility and protection makes their disablement a particularly formidable task. The task is further compounded by the variety of types of armoured vehicles now in service with mechanised armies. The chief target is the heavier and better protected main battle tank which requires special and peculiar consideration if it is to be defeated (vide Chapter 8). Even so, Armoured Personnel Carriers (APCs) and Mechanised (Infantry) Combat Vehicles (MICVs), although provided with less armoured protection than the main battle tank, are nevertheless armoured vehicles and as such require that special consideration must be given to their defeat also.

Structures

Buildings and structures are an obvious type of target but for the purposes of applying firepower to them it is convenient to consider them in three main groups: light, medium, and reinforced structures. The light structures group includes targets such as aircraft hangars, nissen huts and other light storage area constructions, light fortifications and earth works such as those that would be found in a hastily prepared defensive position. Medium structures embraces most conventional buildings and includes the fortifications and earth works that would be found in a well prepared defensive position. The last group, reinforced structures, covers pill-boxes, strongpoints, bridges and other targets constructed in reinforced concrete. Targets in this group are more difficult to defeat than a main battle tank, and this, coupled with the sheer diversity of these targets, makes it uneconomic to design special ammunition to defeat them all. In most cases, therefore, they are considered as secondary targets and when they are engaged, they are fired on with ammunition that has been designed primarily for other purposes.

Equipment

The word equipment when used in this context, as a type of target, covers the whole gamut of material found in the battle area: soft skinned vehicles, parked aircraft on the ground, storage dumps, other logistic installations and facilities, all types of artillery, electronic and communications equipment and all the other many and varied paraphernalia required by military forces in combat. For these types of targets, disablement of the personnel manning these equipments is as important a consideration as the disablement of the equipment, although the primary requirement to defeat the equipment remains valid.

Aircraft in Flight

Aircraft in flight are considered as a separate target category because of their dependence on the environment in which they operate. They also present particular problems of target acquisition and chance of kill (vide Chapter 9). Furthermore, and as with other target categories, the extensive variety of types of aircraft means that the ammunition designer has to further categorise them in order to design ammunition to defeat them. The defeat of fast moving aircraft such as tactical strike aircraft, fighters and fast reconnaissance aircraft requires different considerations to those needed to defeat slow moving aircraft like transport aircraft and others which operate at speeds between 200 and 300 knots, and to those design requirements necessary in ammunition to defeat helicopters.

REQUIRED TARGET EFFECTS

The primary effect required to disable all the types of targets that have been considered so far is a physical effect, capable of causing casualties to people and disruption to other targets. The disablement of people can be achieved by psychological and physiological damage, but - except in the case of nuclear, biological and chemical (NBC) attacks - these effects are difficult to measure and forecast. Furthermore, as psychological and physiological effects of a transient nature are unlikely to achieve any significant level of disablement against well led and well disciplined troops, it is normal practice to discount such effects when assessing ammunition effectiveness.

METHODS OF ACHIEVING TARGET EFFECTS

This volume is concerned only with the conventional methods of attacking targets. NBC warfare is fully covered in Volume IV of "Battlefield Weapons Systems and Technology". So far it has been established that the essential requirement to disable a target by conventional means is to produce a physical disruptive effect at or on it. It follows, therefore, that it is necessary to deliver energy in a usable form to the target or its vicinity and liberate it there to achieve this required physical disruptive effect. Since this energy source must be capable of being delivered to long ranges, then it must also be relatively small and compact. If the energy source is at all delicate or sensitive in its make-up and system of dissipation, it must be protected during launching and delivery. The severity of the constraints these factors impose is governed by the projection system employed and considerable compromise and interplay is thus always necessary to achieve, between the ammunition (the energy source and its means of dissipation) on the one hand, and the gun/launcher (projection system) on the other hand, the correct balance of weight, sensitivity and protection.

Design and Development

There are three principle design factors to be considered in the development of a new weapon system. The first factor concerns the ammunition and whether or not it is technically feasible to provide the type of energy source and energy release pattern required. Once this has been determined the next factor to be considered is the means of projection and delivery of the energy source to the target which must be done accurately and over a reasonable range spread. In practice, the design of a new weapon system often precedes that of the ammunition, or in other instances, new ammunition is required to boost and enhance the performance of an existing weapon. In principle, this approach is wrong and can result in inadequate designs of ammunition being introduced into service. For the best possible target effects to be achieved, and it is the terminal effectiveness of the ammunition which largely determines the overall cost-effectiveness of a weapon system, then the weapon system must be designed around the ideal practical ammunition. Finally, for development and production, the newly designed weapon system and its components must be simple, easy to make and inspect, and also be safe to transport, store and handle. The requirement for simplicity does not preclude advances in design, but is meant to prevent the inclusion of desirable rather than essential features which so often complicate design, increase costs, reduce reliability and delay development and production. For the purposes of this volume, however, it is the first factor which involves the design of the ammunition and its performance that is of concern; weapon systems technology and development are more fully covered in companion volumes in this "Battlefield Weapons: Systems and Technology" series.

Energy Sources Available to the Designer

For the conventional attack of targets there are really only two practical ways of producing the necessary energy at the target to achieve the desired effect. The two sources available are Kinetic Energy (KE) and Chemical Energy (CE). Kinetic energy can be utilised in the form of a solid missile (a "shot") which dissipates its energy on impact and requires no triggering device to control the energy release. Alternatively, chemical energy can be used in the form of a substance, which has considerable potential energy relative to its mass, and which can be released when suitably initiated. Such substances are known, more familiarly, as high explosives and the energy release or initiation mechanism as a fuze.

Both these energy sources can produce the required physical disruptive effect, but which source is better suited to the attack of a particular target depends entirely on the characteristics of the target being considered. For example, against a small vulnerable target, or alternatively against a well protected target requiring considerable penetration to achieve disablement, a kinetic energy missile may give the best chance of success. This may be either because it is the more economic form of attack or because it provides the only method of attack. Against a large lightly protected target, however, where an area effect is required or against a small but strong tough target, which requires high over-pressures to disrupt it, a chemical energy source may be the more appropriate option. Against other targets a combination of effects, using both kinetic energy missiles and the high over-pressures (blast) obtainable from high explosives, may give the required physical effects.

MAJOR TARGET GROUPINGS

It is possible to produce a matching range of ammunition to counter the full range of targets that are likely to be confronted on the battlefield, some of which have been looked at in some detail earlier in this chapter. Clearly, however, it is unacceptable as well as impractical, both economically and logistically, to do this. Instead, the approach used is to establish a list of key targets, in priority, to be defeated and then to design ammunition to give the optimum performance against these targets. A less than ideal or overmatched performance is accepted against the other secondary targets.

Looking at the full spectrum of targets likely to confront land forces, it can be argued that the three main groups or key targets are people, armoured vehicles and aircraft in flight. Whether these three groups are indeed the key targets can be debated, but the point to be made is that having selected these key targets, it is possible to analyse the requirements for defeating each and also to establish the other targets that are likely to be defeated by the same effects.

Attack of People

Man is a delicate and inherently vulnerable piece of machinery. Provided he can be hit, and this is the main problem, he can be physically disrupted with comparative ease and he is also susceptible to pyschological and physiological effects. In the simplest and most cost-effective case an adequate physical effect can be achieved from an energetic missile, relying on its kinetic energy to achieve disablement; the small arms bullet is a prime example. This type of attack is not difficult to mount. Against one man it is best achieved by a single missile, the power of the attack being developed in the projector (the rifle, pistol or other small arm). A group of men can be attacked similarly by a series of single missiles, also developing their power of attack from the projector, as for example in bursts of fire from an automatic or machine gun. Alternatively, a more effective solution against groups of men may be achieved by delivering a chemical energy source, enclosed in a metal projectile, amongst them. On arrival, and when initiated, a proportion of the energy available in the chemical energy source is converted into kinetic energy in the form of a number of missiles, which are obtained from the break up of the surrounding metal projectile case and propelled around the target area. The remaining energy (which it is estimated can be as high as 95%) appears as a local over-pressure in the immediate target area. This combination of missiles and local over-pressure, more usually referred to as fragments and blast, which arises from the initiation of a high explosive encased in a metal projectile, is very useful primarily for the attack of personnel. This combination, however, is equally suitable for the disablement of other types of target, the large blast energy surplus being particularly significant.

Attack of Armour

An armoured fighting vehicle, such as a main battle tank, is a formidable target to the ammunition designer. It is a small, tough, well protected piece of machinery. In addition to its offensive firepower it is specifically designed, by a combination of mobility and protection, to evade and resist attack. To disable it by a physical disruptive effect, therefore, requires a concentrated very high energy attack. To disable it by disabling its crew, either by physical, psychological or physiological effects, still requires a physical effect to defeat its protection first, with sufficient residual energy remaining to achieve the desired effects against the crew. It is not enough just to penetrate an armoured fighting vehicle's protection; sufficient energy must be available, after having penetrated, to do damage behind the protection.

The requirement for a concentrated high energy attack can be achieved directly from either a kinetic energy or a chemical energy source. The essential techniques are covered fully in Chapter 8, but are looked at here to assess their potential against other types of target. The kinetic energy method of attacking armour involves the use of a solid projectile (called a "shot") which is fired at the target at the maximum possible velocity. The chemical energy attack dissipates a high explosive source under controlled conditions at the target. This may be either a shaped charge (or hollow charge) effect, a scabbing effect, or the plate charge effect. In the shaped charge case penetration is achieved by a concentrated jet of very high energy. In the scabbing case "penetration" is achieved by a shock wave reversal detaching "energy", in the form of a metal scab, from the back face of the armoured protection. The plate charge effect achieves penetration by using high explosives to project a metal plate at high velocity to punch a hole through the armour.

Although the kinetic energy shot will also be effective against small tough targets similar to the armoured fighting vehicle, it has limited application against other types of target. The chemical energy forms of attack, however, have considerable potential against other types of target, where their fragments and blast can be used to attack people and material.

Attack of Aircraft in Flight

At first sight an aircraft in flight appears to be a delicate and vulnerable piece of machinery peculiarly dependent on its environment. In fact, in a well designed modern aircraft, its vulnerable area, where its crew and its vital components are, is comparatively small. It is also a difficult target to hit, and in this sense it is its mobility which enhances its protection over and above the protection afforded to it by skilful design. The chance of a kill, given a random hit, against modern aircraft is low. These odds are further increased in the aircraft's favour as the chance of even achieving a random hit is also very low, because of the long flight times taken by gun fired missiles to reach high altitudes. Small kinetic energy projectiles, therefore, have a limited chance of success, unless very heavy, concentrated and wasteful barrages can be brought to bear on a particular aircraft. Thus, kinetic energy missiles really have to be "hitiles" to achieve their effect. Alternatively if a combined kinetic and chemical energy warhead is put in a projectile, it really can become a "missile" in the sense that it does not have to hit to kill. A blast and fragmentation ammunition nature can be effective against an aircraft without actually hitting it.

The concept of destruction, however, is an attainable damage level in this case, as disablement of an aircraft target may lead to its destruction because of its dependence on its environment. The attack of aircraft is covered in Chapters 7 and 9, but again, to allow anti-aircraft warheads to be considered for their effectiveness against other targets, the basic techniques are covered in outline here. Blast and blast/fragmenting munitions are used, as well as rather more sophisticated warheads containing special kinetic energy missiles. These special missiles derive their energy in the target area from a chemical energy source, giving a linear or specific area concentration of kinetic energy (for example the continuous rod warhead; see Chapter 9). All these warheads have a potential against other forms of target which are susceptible to blast and fragments.

SUMMARY

It is technically feasible to design a spectrum of warheads for the ideal engagement of a particular type of target. Economic and logistic considerations, however, make this a totally impractical and unacceptable option to adopt. A diversity of types of ammunition and warheads should be restricted to an absolute minimum, commensurate with the need to preserve a high chance of success against all targets. An analysis of target frequency distributions and the problems peculiar to the disablement of certain targets shows that there are three key targets to consider; people, armoured vehicles and aircraft in flight. Ammunition designed to disable people and armoured vehicles are between them suitable for the disablement of most other types of target including, in some instances, aircraft in flight. The basic concepts introduced in this chapter are common to all types of targets and they are further developed in subsequent chapters which cover particular targets in more detail.

SELF TEST QUESTIONS

QUESTION 1 What is the overall kill chance expression, and why is it necessary to have such an expression?

Answer ....

....

....

....

QUESTION 2 Give the damage levels usually considered by the ammunition designer.

Answer ....

....

....

....

QUESTION 3 Which damage level does the ammunition designer primarily aim to achieve, and why?

Answer ....

....

....

....

QUESTION 4 List the main types of targets likely to be confronted by land forces.

Answer ....

....

....

....

QUESTION 5 Are psychological and physiological effects relevant considerations when assessing ammunition effectiveness?

Answer ....

. . . .

. . . .

. . . .

QUESTION 6 What are the principal design factors to be considered in the development of a new weapon system?

Answer . . . .

. . . .

. . . .

. . . .

QUESTION 7 What are the main energy sources available for the conventional attack of targets?

Answer . . . .

. . . .

. . . .

. . . .

QUESTION 8 What are the main requirements for ammunition designed to attack people?

Answer . . . .

. . . .

. . . .

. . . .

QUESTION 9 Describe the main requirements for anti-armour ammunition.

Answer . . . .

. . . .

. . . .

. . . .

QUESTION 10 What are the main considerations involved in designing ammunition to attack aircraft in flight?

Answer ....

....

....

....

ANSWERS ON PAGE 248

3

The Attack of Personnel

INTRODUCTION

Most modern weapons used for the attack of personnel are designed to transfer a quantity of energy as rapidly as possible to the target, in order to damage it in some way, either physically or psychologically. Physical effects result in casualties and damage to material, whilst psychological effects prevent the enemy from using his weapons effectively and deprive him of his will to fight. It takes no more energy to do this than it did in the stone age. Before explosives and firearms were known, attack of personnel was by stone or lead slug propelled by hand or sling. Firearms arrived in 1250 when the Moors adapted cast iron buckets to project stones using a powder charge. Gunpowder and cannon improved matters from 1350 onwards but rifled barrels (for accuracy), although known in 1525, were not used in combat until the 30 years War (1618-48). The introduction of steel to replace iron, together with high explosives (HE) and propellant to replace gunpowder, at the turn of the 18th century, improved matters considerably. Perhaps this is why in the 20th century so far nearly 100 million people have died in well over 100 wars!

Initially there was very little science applied to the subject and often devices were either quite useless or produced an overkill. Many theories have been put forward regarding the energy required to produce a casualty, and several methods of optimising such effects and ways of assessing target damage have been used. Some of these are discussed later in this chapter. It is convenient now to describe the target and explain the complex problem in assessing target effects.

THE TARGET

The human being is a relatively small and complex target measuring about $0.42m^2$. It may be protected in some way by clothing, flak jacket, steel helmet and visor; it may be in the open or behind cover; it may be standing, lying, moving or in some other position. It can also be fit or unfit; alert or tired; well motivated or depressed. There are many other aspects of human behaviour and attitude which may affect it as a target. It has bone, muscle, nerves, arteries and obvious vulnerable areas. The target can be affected by blast or fragments or a combination of both. It is this complexity that makes assessment extremely difficult.

Assessment has taken many forms over the years but the current method is to assume the target is standing palms forward facing the attack, the frame being identified through 108 slices taken horizontally and a wound tract being applied in each of six 60° paths through each tract. Medical assessment is then applied which takes into account the structure and this is extended to include the influence of motivation, battle effects, posture etc. This provides the basic information for target effect assessment and is ideally calculated by computer. The design of all modes of attack of personnel must have these aspects as a prime objective.

MODES OF ATTACK

General

The most efficient mode of attack requires a high chance of a hit, high $mv^3/2$, rapid energy transfer and no unnecessary overkill. The first is obvious whilst the second shows the importance of velocity. The third should ideally be the complete transfer of energy instantaneously to the target and the fourth relates to economy or efficiency of the system.

The common method of transferring energy to a target is by fragment attack, the fragments being produced by a HE projectile, mortar bomb, grenade or other projectile. Bullets and flechettes are not normally classed as fragments but they are projectiles in their own right, bullets being generally more selective.

There is an optimum size fragment which is used by designers but it is clear that this can only be a very general statement of the requirement because of the variables involved in the target which include vulnerability, protection, distance travelled by fragment, HE back up, material used and so on.

Fragment Producers

Fragment producers have received considerable attention since the days of cannon balls filled with gunpowder or projectiles filled with steel balls, although similar projectiles to the 'shotgun' type are now used in counter ambush weapons. These are known as canister rounds and are dealt with later. Modern HE shells produce fragments of various sizes and shapes - some an overkill, some too small to harm anyone or anything but the majority being within the optimum fragment size requirement for that particular projectile. This method of producing fragments is known as natural fragmentation. An example is shown in Fig. 1. It will be readily seen that where the ratio of high explosive to metal is high, fragments will be of the optimum size, but at the nose and base of the projectile where the ratio is low, larger fragments will be produced.

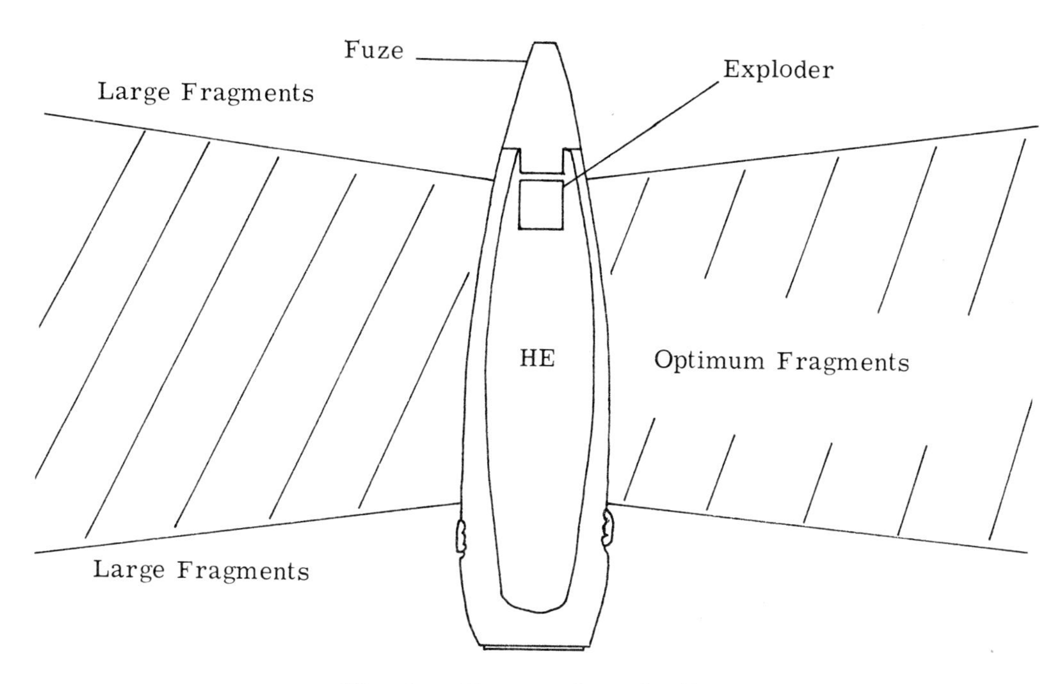

Fig. 1. Fragment production

The tendency now, particularly with mortar bombs and grenades is to opt for pre-notched containers or pre-formed fragments held together by resin and enclosed in a thin metal container. Although these designs provide the most cost effective fragment producers there are launch restrictions, which prevent their use at present in artillery weapons where stresses on the projectile during firing are large.

Bullets

Bullets are designed primarily for the attack of personnel and are generally selective rather than random as for HE shell fragments. The early bullets were fat, heavy and slow, measuring 10-15 mm in calibre and consisted usually of lead which deformed on hitting the target. Such material left considerable fouling in the weapon. The calibre continued to shrink over the years through 9 mm and 8 mm (.303 in) down to the 5.56 mm which seems now to be the popular calibre.

Flechettes

Flechettes are a fairly recent addition to the anti-personnel armoury and usually exist as bundles of tightly packed small darts in a projectile being released during flight by a small charge and which result in a mass attack effect from the air. Known as beehive shell (because of the honeycomb effect of the tightly packed darts) they are mentioned again in Chapter 10. Attempts have also been made to put two or three flechettes into a single bullet skin with some success.

Fragment Characteristics

The most significant factors which affect the performance of a fragment and thus its incapacitation potential are: mass (Fig. 2) velocity (Fig. 3) and shape. Whilst mass and velocity are normally associated in the kinetic energy form $\frac{1}{2}mv^2$ or momentum mv there have been many interpretations of their actual value in wound ballistics. Values have ranged from $m^{0.4}v$ through mv^3 to $mv^3/2$ and this is the form currently accepted. The reduced effect of velocity from the normal kinetic energy value is due to the human frame being considered non-rigid as compared to a hard target. Shape however is not quite so straightforward because its effect on the target depends very much on its orientation to the target as it makes contact and also its mode of travel through the air. A heavy fragment will have greater carrying power but may not have the right shape to penetrate. Steel balls maintain their velocity better than irregularly shaped fragments or deformed lead balls. A long thin bullet or flechette may become unstable at certain ranges and lose velocity quickly before reaching the target. On the other hand, it may have too much velocity, because it is less affected by air resistance, and go right through the target without inflicting incapacitation. The final design is a compromise of all the above characteristics to give the best possible anti-personnel device.

Energy Transfer

Casualties are caused by the transfer of energy from the fragment to the target. Severity of the casualty normally depends on the amount and rate of transfer of the energy. The rate of transfer can be enhanced by marginally stable fragments giving up all their energy to the target or by a fragment which is deflected in its passage through the target. If energy transfer were small or the rate of transfer is low then a mild casualty would be the result but this assumes that no vital organ or artery is damaged in the process. The high velocity stable fragment, such as the 7.62 mm bullet at ranges of 200-800 m, gives up little energy at the target as it passes through unless inhibited within the target. On the other hand when the energy transfer is large or the rate of transfer is rapid a much more severe wound is inflicted. Ideally the requirement is for a high velocity small profile fragment which is not degraded by air resistance yet can transfer all its energy rapidly at the target. There is a minimum required weight to allow the fragment to carry to the target accurately. The early dum-dum bullet, produced in India by cutting the tip off the bullet to give a wider cross section area on impact and thus increasing its 'stopping power', was an attempt to try and improve the rate of energy transfer. Because of its effect the Hague Convention outlawed its use in the anti-personnel role. The term 'stopping power' or 'knock down' ability has been quoted in many publications over the years and a value of 80 Joules has been associated with it. This value is not considered meaningful as there are many factors to be considered in any one situation such as impact velocity, tumble possibility, yaw angle on strike, density and attitude of target and so on. To quote a figure of 80 Joules as an all encompassing figure appears to have no biological or physical basis.

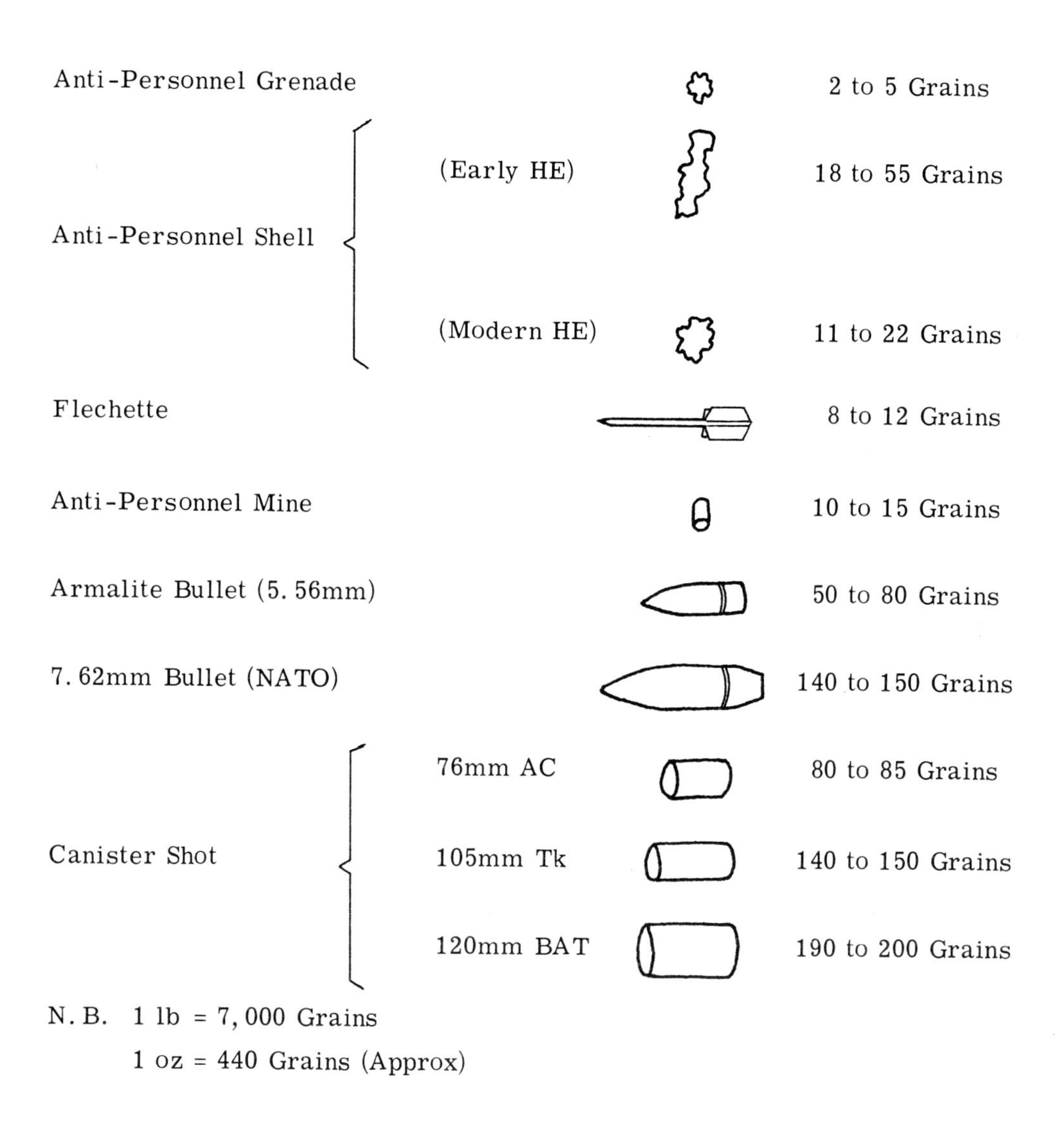

Fig. 2. Typical fragment masses

FRAGMENT VELOCITIES (V_F)

The graph below shows the order of fragment velocities obtained with variation in d/t ratio which is the amount of explosive behind the case thickness. This ratio varies along the length of the device and must not be confused with the Charge to Weight Ratio (CWR) which is the ratio of the total weight of the explosive to the total weight of the device. Two examples are given: TNT/steel (old shell) and RDX/TNT/Improved Steel (modern shell) which consists of an improved HE filling with a higher tensile steel case.

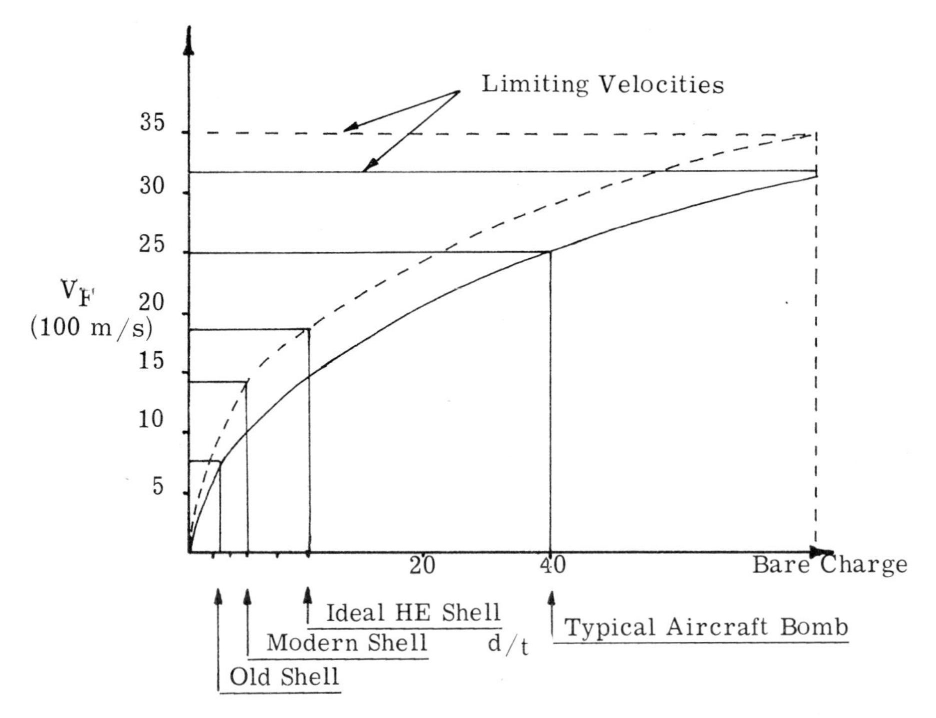

Fig. 3. Fragment velocity/variation in d/t ratio graph

Notes:

1. Although the aircraft bomb gives a high value of V_F the fragments generally consist of a few large sections of the bomb wall with poor ballistic shape and giving no predictable fragment pattern.

2. If the predicted ideal figure for a conventional shell were obtainable, fragment velocities of the order of 1800 m/s would be expected.

3. In fragmenting warheads for guided missiles, where the launch conditions are low compared to guns, it is possible to obtain fragment velocities up to 2,400 m/s because the strength of the case can be much less.

Battlefield Wounding Criteria

A fragmenting device should ideally produce as many lethal fragments as possible to make the target incapable as early as possible. A device which meets this requirement is said to have a good 'lethality' which is a term not easy to quantify but is an indicator to a system's effectiveness, particularly with regard to lethal areas. The effect produced by a fragment will vary with the condition of the target as explained earlier. To include this in an assessment of casualties is obviously necessary and with this in view a number of typical battlefield wounding criteria have been specified and, against these, the chance of a kill given a hit can be assessed. The probability of a kill given a hit is generally written as P_{hk}.

The most severe criterion requires the incapacitation within 30 seconds of the strike of a determined well protected target occupying a good defensive position. This is known as the 'Defence 30 sec Criterion'. It should be noted that in this situation the chance of a hit is low and the random hit is likely to be on the head and shoulders which may be protected by steel helmet, visor and flak jacket.

Another criterion is the 'Assault 5 minutes Criterion'. This requires the incapacitation of a highly motivated assaulting infantry man in 5 minutes. In this case the chance of a hit is greater and the overall chance of a kill is also greater. Originally there were 14 criteria but now about 4 only are used for assessment.

The assessment is complex and time consuming and usually requires the services of a computer. For each battlefield wounding criterion the results are plotted to give the variation of kill chance against fragment mass and velocity (m and v) as shown below in Fig. 4.

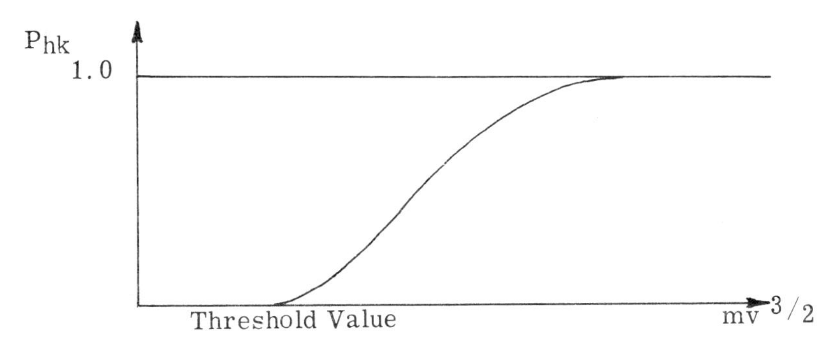

Fig. 4. A typical curve of P_{hk} against $mv^3/2$

The best equation that can be fitted to this curve is of the form:

$$P_{hk} = 1 - e^{-a(mv^3/2 - b)^n}$$

where m = mass, v = velocity and a, b and n are constants associated with the human stress situation and wounding time dictated by the criterion being considered.

The Performance of a Complete Fragmenting Device

The performance of a complete fragmenting device is influenced by several factors, some of which are now discussed. The analysis of performance is again complex and is not simply the sum of the performance of the individual fragments.

In the fragmentation device the HE filling is initiated by a fuze and at the point of initiation a detonation wave builds up and is propagated through the filling. The casing in the rear of the wave front is accelerated rapidly outwards by the pressure of the detonation. Expansion outwards is accompanied by a reduction in wall thickness until the case breaks up (Fig. 5).

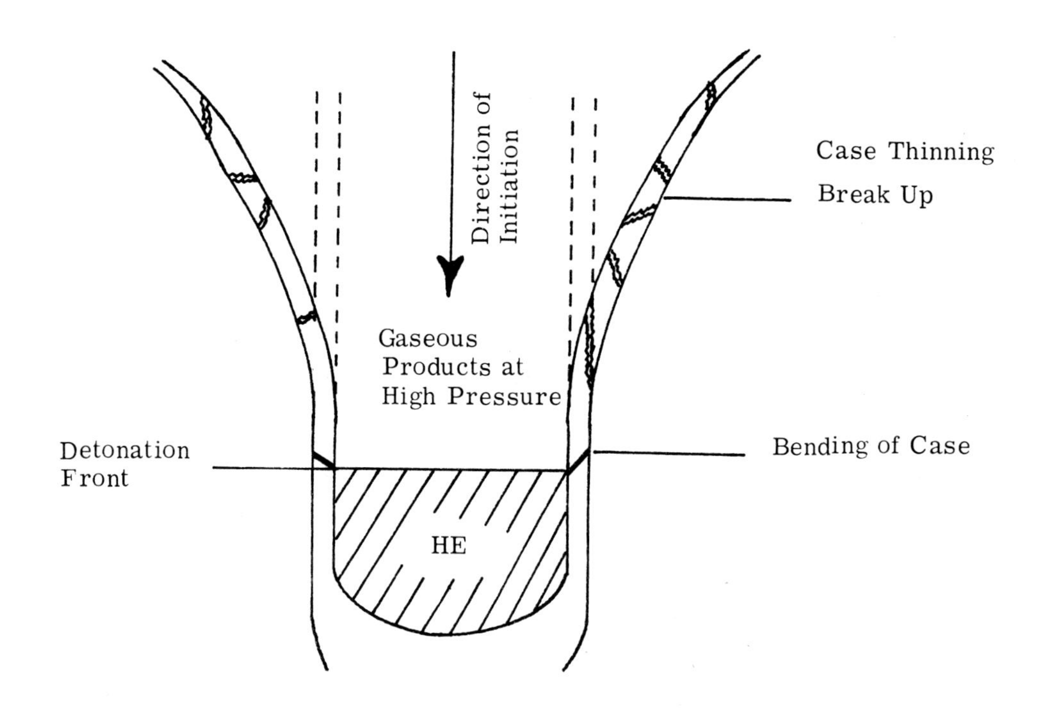

Fig. 5. Fragment formation

Spatial distribution of fragments. The design profile of the device and the variation in explosive/casing ratio affect the fragment distribution. If the device is a projectile the spin rate will also influence the distribution but not the overall pattern. The distribution is usually symmetrical about the longitudinal axis. When the factors of remaining velocity and angle of arrival are considered the pattern is distorted. The use of notching techniques, or pre-formed fragments, tends to produce a more regular distribution but the pattern is still affected by velocity and angle of arrival.

Orientation of device to target. With a conventional high explosive projectile the most effective distribution of fragments is attained when the axis of the projectile is vertical. As the axis moves away from the vertical an increasing number of fragments are projected into the air and fall relatively harmlessly under gravity whilst others are lost into the ground.

Height of burst. The height of burst has a significant effect on the performance of a fragmenting device and is achieved by the fuzing system. Ground burst effects depend on the type of ground and climatic conditions as well as sensitivity of the fuze action. Proximity fuzes are used to give air burst effects although some time mechanical fuzes may be encountered which give a less precise result. A burst height of 2 to 4 metres is ideal for anti-personnel effect. Figure 6 overleaf shows the effect of ground burst, airburst and angle of impact on the lethal area.

Average exposed area of target. The following aspects should be considered: attitude of the target ie is the man standing, kneeling or prone; type of ground in which the target is situated and the protection afforded by his clothing and equipment.

TESTING OF FRAGMENTING DEVICES

It is difficult to be precise regarding the lethality of fragmenting devices because of the variations in the target and the conditions surrounding it. Nevertheless there is a mathematical model for this and also a series of static tests to determine the fragment distribution. Such tests in the United Kingdom are carried out by the Directorate of Proof and Experimental Establishments and are very time consuming. The device is usually mounted about one metre from the ground with its major axis horizontal. Screens are arranged around the device so that when the device is detonated the fragments are retained by them. The number and depths of strikes indicate distribution and velocity. Velocity screens on which the velocity can be deduced from fragment penetration are also provided for greater accuracy in measuring velocity. From the fragment distribution and velocities obtained and analysed, models can be built by feeding in variables such as angle of arrival and remaining velocity of device together with height of burst. By applying air resistance factors to fragments an assessment of the chance of a kill given a hit can be made in an area around the point of burst. This area is referred to as the lethal area and provides the designer and user with the information regarding the effectiveness of the device.

CONCLUSION

A well tried, effective and practical technique for attacking personnel is by a combined blast and fragmenting device. Most casualties are caused by the fragments but blast has a psychological effect. The modern tendency is to keep the fragment mass low and velocity high. This gives an adequate density of fragments within the required radius of effect, each fragment being capable of causing severe wounds. There has been a general improvement in the performance of anti-personnel devices in recent years and further improvements can be expected as new techniques are developed. A number of conflicting factors confront the designer and inevitably, as with many other technical problems, the result is usually a compromise.

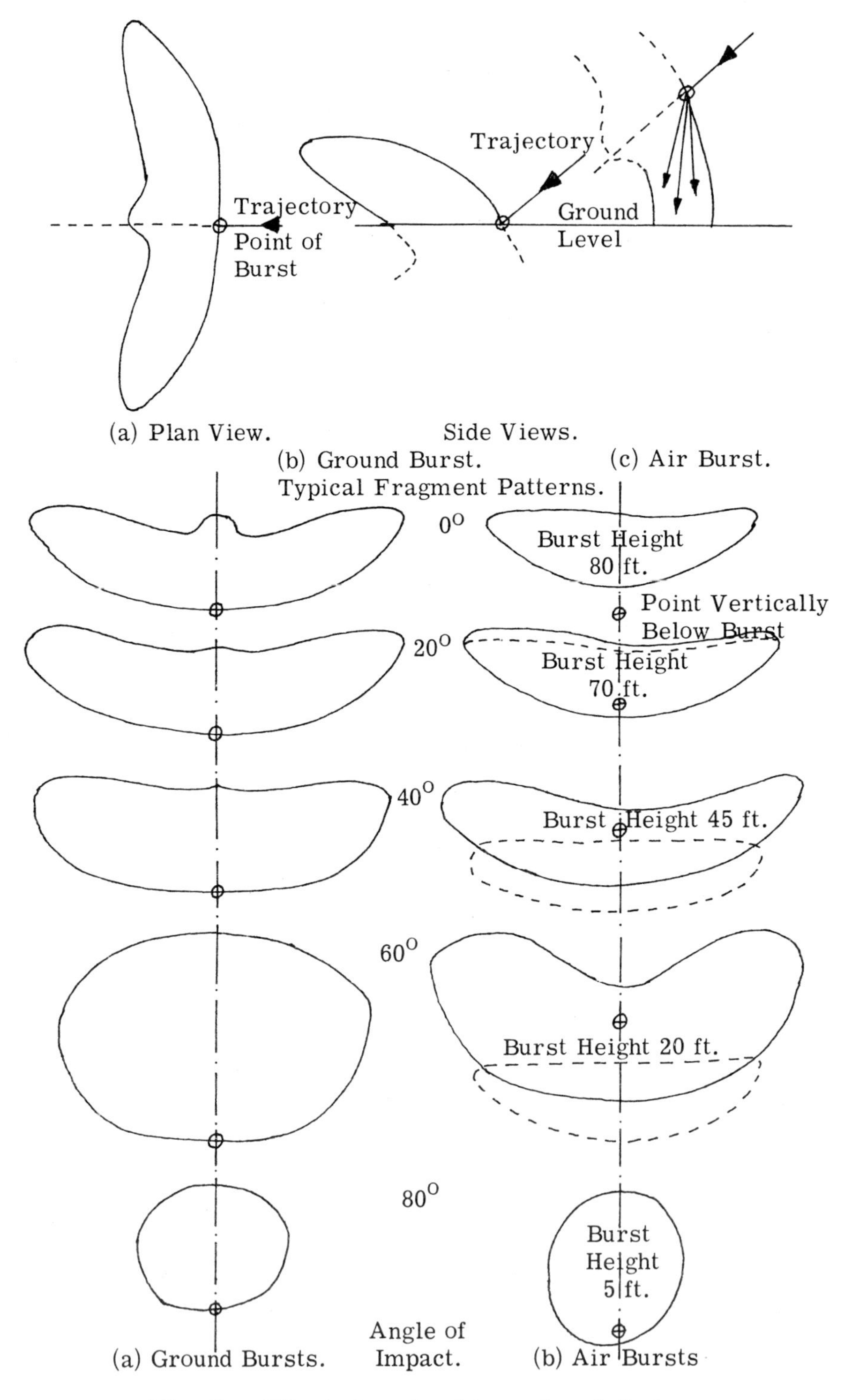

Fig. 6. Effect of angle of impact on lethal areas

SELF TEST QUESTIONS

QUESTION 1 Describe the various aspects of a human being as a target.

Answer ....

....

....

QUESTION 2 How is a human target assessed?

Answer ....

....

QUESTION 3 What are the essential requirements for the defeat of personnel?

Answer ....

....

QUESTION 4 List the factors which affect the performance of a fragment.

Answer ....

....

QUESTION 5 What is meant by lethality?

Answer ....

....

QUESTION 6 Explain "Defence 30 sec Criterion".

Answer ....

....

QUESTION 7 Describe briefly the action of a fragmenting device.

Answer ....

....

QUESTION 8 How does the angle of arrival affect the fragmentation pattern?

Answer ....

....

QUESTION 9 Why is an airburst device more efficient than a ground burst?

Answer ....

....

QUESTION 10 Is the psychological effect of a fragmenting device as important or significant as the physical effect?

Answer ....

....

ANSWERS ON PAGE 249

4

Explosives and Propellants

EXPLOSIVES

Introduction

No publication on ammunition can be complete without some basic input regarding explosives. It is important to understand the characteristics of explosives, their limitations and in particular the framework within which they are controlled from the safety aspects.

History

The first explosive, gunpowder, was discovered by the Chinese and reached Europe about the 12th century. It was, at first, only used for propelling primitive rockets and was called blackpowder. Although hygroscopic it was the only explosive available for many years and was improved by Roger Bacon and others, taking various mixtures of saltpetre, charcoal and sulphur from 40:21:21 to 75:15:10 which is the current mix. Gunpowder is a mechanical mixture where carbon (charcoal) and, to a lesser extent, sulphur are the fuels and potassium nitrate (saltpetre), the oxidant.

With the introduction of cannon it was found that gunpowder gave erratic performance and this prompted a scientific approach to the mixing and preparation processes. It was found that the rate of burning of gunpowder, a vital requirement for the ballistician, depended not only on the mix but on the size of the grain and its regularity. The process of "corning" which produced grains of a predetermined size and "moderating" which was carried out by polishing the grains with graphite were introduced, and by the end of the 16th century manufacturing processes were standardised. Modern gunpowder is referred to by its grain size (eg G12). The larger the number the smaller the grain. It is still used today in many ammunition components.

Considerable effort was made in many countries to try and find another explosive without much success. Even Pepys records an experiment where he heated a few grains of 'fulminating gold' (the size of a match head) in a silver spoon and blew a hole in it. The 19th century showed a real advance when the effects of nitrating organic substances were discovered. Laurent (1841) obtained a high explosive (HE) Picric Acid from the nitration of phenol. Schoenbein (1845) nitrated cotton finding guncotton and later produced a propellant known as Poudre B from nitro-cellulose (NC). The French used this at the turn of the century. (The reader should be aware at this stage that gunpowder has a very low performance compared with a modern high explosive and is usually classed as a propellant which is designed to burn rather than detonate, a requirement of a HE). The nitration of glycerine (Sobrero, 1846) produced nitro-glycerine (NG) and from this (Nobel, 1867) produced dynamite. Later, Nobel, using NG and NC produced a thin horny sheet which, when cut, was a good propellant called ballistite. After this, experiments were carried out at Woolwich in which NC and NG were incorporated in the presence of a solvent, acetone,and this product was able to be extruded in long cords which, when dried, became a hard brittle propellant called cordite. This propellant was approved for service use as a propellant in 1871 and was the forerunner of a wide variety of propellants now used.

Finally, the nitration of toluene (1867) led to tri-nitro-toluene (TNT) which has been the main British HE since World War I and is used very widely still.

Classification of Explosives

For convenience and usage it is necessary to understand the basic difference between high explosives and propellants. Figure 1 shows a complete round of ammunition which is to be loaded into the gun and fired. In so doing the firer starts a chain of events which results in the projectile leaving the gun at high velocity carrying its payload (in this case HE) with it to the target. This process consists of the initiation of a small amount of sensitive mixture in a cap which in turn is boosted by a magazine to ensure instantaneous ignition of the propellant which burns very rapidly (about a few hundred metres per second), producing a high temperature and high volume of gas which propels the projectile out of the bore. When the HE projectile reaches the target it is initiated again by a small amount of a sensitive mixture in a detonator located in the fuze, which in turn is boosted by the fuze magazine and exploder pellet to the main filling of HE, which detonates up to 9000 metres per second and breaks the steel shell into fragments. The two processes are referred to as explosive trains.

It will be seen that whilst the mixtures in the cap and detonator are similar they are designed to function in two distinctly different ways. The cap provides heat or flame which produces a rapid burning of the propellant but the detonator is the starting point for a detonation wave or molecular disruption of the HE which detonates. In short, propellants burn rapidly or deflagrate but HE detonates. The table at Fig. 2 gives some distinctions between burning and detonation.

To complete the picture there is a further group of burning explosives which are referred to as pyrotechnics and these can be loosely termed 'fireworks'. Details of pyrotechnics are given in Chapter 16.

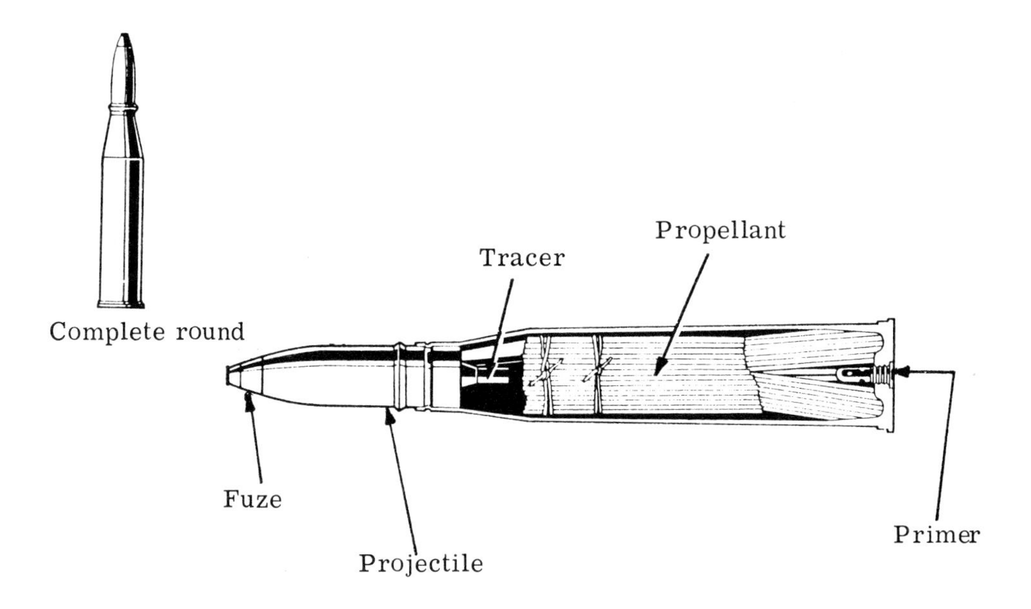

Fig. 1. Typical fixed ammunition

Control of Explosives

Explosives constitute risks to society therefore regulations are needed for the protection of all. Explosive Acts of 1875 and 1923 confer on the Home Office this responsibility which is exercised through the Explosives, Storage and Transport Committee (ESTC). Aspects such as manufacture, movement, storage, handling and so on are strictly controlled and any new explosives become authorised only when fully tested, they are then added to the Home Office list. The Health and Safety at Work Act 1974 also has a bearing on explosives.

INITIATORY EXPLOSIVES

Initiatory explosives are explosive train starters and can be made to function reliably from a small mechanical, electrical or thermal stimulus. There are two basic types, igniferous and disruptive. Those which ignite some other substance are called igniferous and those which produce a shock pressure or cause detonation are called disruptive. They are usually filled with one or more of the following explosives in layer form: lead styphnate, lead azide, tetryl, etc. It should be noted that an igniferous train comprises only igniferous elements, terminating in a propellant or pyrotechnic charge. On the other hand, a disruptive train, although terminating in a HE charge, may commence with either a disruptive or igniferous initiator.

BURNING	DETONATION
1. All explosives burn at the outset when initiated by igniferous means. (Exceptions: water-based compositions).	1. Most explosives are capable of detonation if sufficient stimulus is applied.
2. Burning is slow compared with detonation. Burning rates lie between 0.001 m/s and 500 m/s.	2. Detonation is much faster than burning. Velocities of detonation lie between 1800 m/s and 9000 m/s for solid explosives.
3. Because of the comparatively low linear rate and because of conductive and radiative factors, there is often a tendency for the flame to spread along unburnt surfaces faster than it passes into the bulk of the explosive. Burning is a surface reaction.	3. The mechanism of detonation causes propagation to occur throughout the bulk of the explosive, proceeding radially from the point of initiation within the charge. The surface of the charge is normally reached from within by the wave, and is merely the boundary at which the wave ceases to be self-supporting. Detonation is not a surface reaction, it is a shockwave mechanism.
4. The linear rate of burning increases with ambient pressure.	4. The velocity of detonation has a limiting value for a given explosive. It is virtually independent of ambient pressure.
5. The linear rate of burning is not dependent on the size of the burning charge. There is virtually no critical diameter effect.	5. The velocity of detonation is dependent on charge diameter (for small charges). Detonation in a linear charge fails below a critical diameter of significant dimension.
6. Burning is initiated by direct heat or flame, not usually by explosive shock. Burning may convert to detonation if conditions are favourable.	6. Detonation is initiated by shock or by transition from burning. It does not usually revert to burning; if propagation fails, the charge remains chemically unchanged.
7. Burning does not, intrinsically, create reports or shock phenomena, but noise and airblast may occur when cumulative overpressure is vented into the atmosphere.	7. Detonation creates a report which is caused by the shockwave emerging from the charge into the surrounding air.

Fig. 2. Some distinctions between explosive burning and detonation

Requirements of an Initiator

The most important of these is high sensitivity with safety but initiators must also be extremely reliable and consistent in their reaction time. Clearly, when firing a gun or mortar, speedy reaction is necessary. This is vital where the target is moving, so slow initiation is unacceptable.

SECONDARY EXPLOSIVES

Uses

Secondary explosives have a high energy output but cannot normally be detonated as easily as initiatory explosives. They are used as main fillings for projectiles, bombs and torpedoes and comprise the main component for demolition explosives. They need an explosive train to detonate them, as already explained, the speed at which the shock wave travels through the explosive depends on the density of loading. Most explosives are capable of detonation when initiated in a confined state and the result is a breakdown of the substance into gaseous products.

Characteristics of Detonation

The process of detonation produces a shock wave which is supersonic in the explosive and its velocity has a limiting value which is characteristic of a particular explosive. This limiting shock velocity is the Velocity of Detonation (V of D) and always exceeds the velocity of sound in the explosive. The velocity of sound in air is 330 metres per second but in crystalline explosives it is of the order of 3000 metres per second.

A detonation shock wave exerts an extremely high pressure in the shock front and this pressure, called detonation pressure, can be as high as 15000-30000 MNm^{-2}. This pressure causes the projectile to expand and break up. The detonation wave from an isolated spherical charge initiated from the centre is spherical in shape. Waves can be shaped by explosive 'lenses' to yield planar or convergent waves.

Requirements of Explosives

Because explosives are used to create damage and inflict casualties they are a potential hazard, they must therefore be stable over long storage periods and have a sound shelf life. Explosives are used in conjunction with various other materials in ammunition and they should be used only with compatible materials. They should be comparatively insensitive to impact, friction, electrical discharge, heat and shock but also need to be reliably initiated when assembled with an explosive train. The usual requirements of cost effectiveness, availability and handleability must also be borne in mind.

Effectiveness of Explosives

The effectiveness of explosives depends on the amount of energy available per unit volume and the rate of release of this energy when detonated. It will be appreciated that a rapid release of energy is required in a HE projectile to produce lethal fragments, but a slower release is needed when blasting or mining with explosives. To measure effectiveness various parameters are used and these are influenced by the chemistry of the explosive i.e. its chemical composition and structure. Some of the parameters employed are similar to those mentioned above under characteristics of detonation, but there are others such as Power Index and Brisance or shattering effect. The power of an explosive is the work done or energy released during the event. It is usually expressed as a percentage compared with Picric Acid (an early explosive) and the index is a figure as shown below in the table at Fig. 3. Shown also in the table are velocities of detonation (V of D) and values of figures of insensitiveness (F of I) which relate to the sensitiveness of an explosive as a measure of its reaction to certain external stimuli eg impact and friction.

Explosive	Power Index	V of D m/s	F of I	Type
Lead Azide	37	4-5000	10	Initiating
Tetrazine	40-50		8-13	Initiating
CE (Tetryl)	122	7500	70-75	Intermediary*
Picric Acid	100	7250	100	Secondary
TNT	95	6950	115	Secondary
RDX	167	8400	60-65	Secondary
PETN	120	7450	85	Secondary

*The term Intermediary is not now used but does show the second item in an explosive train between the initiatory and secondary explosives. Intermediaries are now classed as secondary explosives.

Fig. 3. Some explosives values

Choice of Explosive Filling

Selection of an explosive filling is usually a compromise and depends very much on its treatment during firing or launch, its required target effects and its reaction with the target. Forces acting on a filled shell fired from a high performance gun are very high compared with those in a bomb or torpedo. On the other hand the HE in a high explosive squash head (HESH) round must not detonate on impact with the target until initiated by the fuze from the rear. Generally, HE with a F of I of less than 100 are not normally used on their own as main fillings in the British service, therefore explosives such as RDX, HMX and PETN must be suitably desensitised with TNT, wax or oil. These mixes result in some loss of power and reduction in V of D but enable the filling to be safely launched. The addition of aluminium increases the power of an explosive but decreases V of D. This slower release of energy is utilised in Torpex for underwater weapons and blasting explosives.

Methods of Filling

The methods used for filling high explosives into shells, bombs etc. include casting, pressing, extrusion and handfilling, with vibration as necessary. Casting is the process of pouring molten explosive (80-85°C) into the store and allowing it to cool and solidify under carefully controlled conditions. TNT, RDX/TNT, RDX/Wax and Torpex are among those fillings that can be handled by this method. A convenient method for high output rate of small and medium calibre shell is press filling, where power operated presses are used to press a number of increments into the store at pressures of 90-150 MNm^{-2} to ensure correct and even density. Extrusion filling has limited use but can be suitable for thermoplastic explosives. Hand stemming used mainly for Amatols was improved by introducing a vibrating metal probe but is not now in general use.

Future Trends

There is not likely to be any marked increase in the performance of explosives in the future because the amount of energy available per unit volume is nearing the maximum. Modern explosives have to withstand increased stresses and temperatures whilst remaining stable and reliable. Plastic bonded explosives are becoming increasingly popular because they fulfil modern requirements for high velocity and heat resistance.

PROPELLANTS

Introduction

Modern propellants have been developed from cordite, first produced in Woolwich in the late 19th century where it met the requirements of the day. Since then a wide variety of propellants and additives have been developed and used.

Propellant Requirements

A propellant must give consistent performance to ensure that when a series of projectiles are fired or missiles launched the velocities show as little variation as possible. This means that a particular mix of propellant - referred to as a lot - should be homogenous and its production carefully controlled. Ideally a propellant should be smokeless and flashless in order that the firing position remains undetected, but it is not possible to produce such a propellant. Flashless propellants tend to give a smoke effect and vice versa. It would not be convenient from the user's point of view to have smokeless propellants for day use and flashless for night use. Due to the high temperatures involved in propellant burning, erosive wear of the barrel is inevitable but several methods have been introduced in order to reduce the erosive effects. Propellants must be stable in storage and should stand up to cyclic temperatures throughout their shelf life. Modern propellants remain stable for many years given reasonable storage conditions and this is due to various additives which are mixed with the propellant during manufacture.

Propellant Additives

These can be grouped as gelatinisers, plasticisers, stabilisers, coolants, moderants, flash inhibitors, surface lubricants and anti-wear additives. To provide homogeneity, coherence, strength and flexibility organic solvents and NG are used as typical gelatinising agents whilst organic esters are used as typical plasticising agents. Stabilisers such as carbamite and diphenylamine are used to increase the shelf life by preventing the chemical breakdown. Coolants are non-explosive ingredients such as carbamite and oxamide, which absorb heat when the propellant burns and cools it down. Moderants comprise substances such as carbamite and graphite which are used to form surface coatings on propellant grains to reduce the burning rate early in combustion. Flash inhibitors provide nitrogen to dilute muzzle gases; the main one in use is picrite (nitroguanidine) and potassium cryolite is a catalytic after-burn inhibitor. Graphite provides a good surface lubricant for granular propellant and also reduces pick-up of static charges. To prevent high rate of wear in the gun, polyurethane liners, Swedish additive and combustible cartridge cases have been used. Modern propellants will contain one or more of these additives depending on their use.

Propellant Types

Single base propellants contain nitrocellulose as the main ingredient, double base have approximately equal quantities of nitrocellulose and nitroglycerine whilst triple base types have about one quarter each of nitrocellulose and nitroglycerine and half picrite . The basic differences of 3 typical propellants are shown in the table below at Fig. 4.

% Composition	Single Base	Double Base	Triple Base
Nitro-cellulose	84	49.5	21
Nitro-glycerine	-	41.5	21
Picrite	-	-	55
Moderant	10	-	-
Stabiliser	1	9	3
Plasticiser	5	-	-

Fig. 4 Propellant types

Rate of Burning

Propellants burn (deflagrate) smoothly over a wide pressure range and the rate of burning depends on the chemical composition, shape and surface area. As pressure increases so does the rate of burning. A typical burning curve is shown in Fig. 5 and is related to a gun propellant.

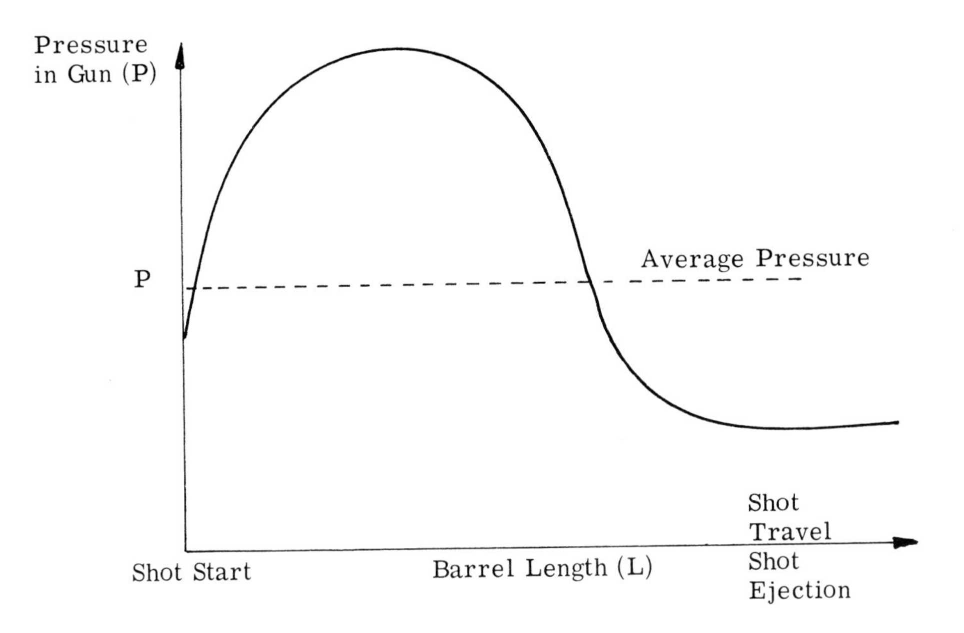

Fig. 5 Typical burning curve

The higher the pressure the stronger the gun barrel and chamber needed to contain the pressure. Piobert's Law (1839) shows that burning is by parallel layers (Fig. 6) and the burning rate was linked to pressure by de Saint Robert's Rule (1862). These and other laws enable the ballistician to calculate the internal effects of a particular quantity, type and shape of propellant. There are many shapes and sizes of propellant grain. The term grain applies to a single piece of propellant whether it be a tiny stick in a small arms cartridge or a large one in a rocket motor. Examples of various shapes are shown overleaf Fig. 7. The size is the least dimension to burn through thus; a solid cord will be completely consumed when its diameter has burned through. The size influences the maximum pressure and time when 'all burnt' is reached, whilst the shape dictates the way in which the surface area varies during burning. Some other aspects of propellant burning are: the Force Constant or Force which is obtained by burning 1gm of the particular propellant in 1cc; Vivacity, the rate at which propellant gives up its energy; and Quickness which is Force x Vivacity.

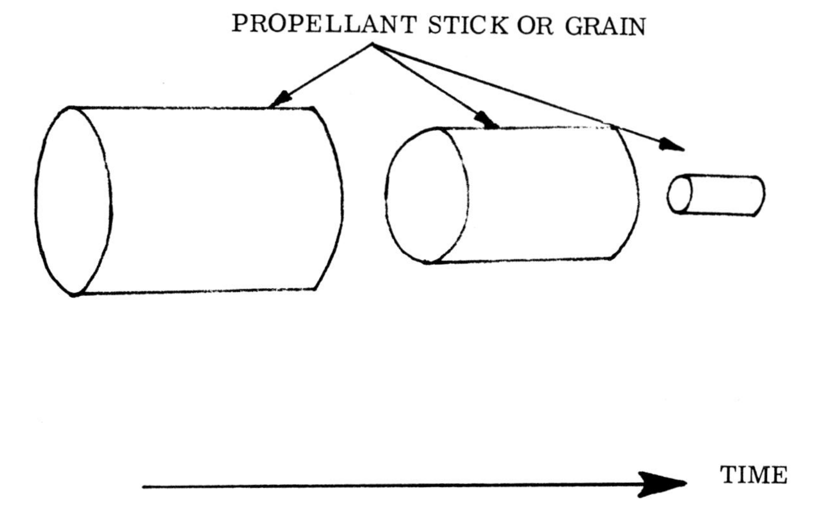

Fig. 6. Burning by parallel layers

Manufacture of Propellant

There are three basic methods as outlined below.

Solvent method. For single base propellants NC is introduced in wet form, dehydrated with alcohol and gelatinised with ether. The resultant dough is then extruded through dies, chopped and dried. For double and triple base propellants the NC and NG is mixed under water, dried to a paste which is then formed into dough by mixing the other ingredients as appropriate in an incorporator with the solvent (acetone and water). The dough is then pressed and extruded to the required shape. The final drying process can take from days to weeks and thus requires much space and time.

Semi-solvent method. This is similar to the solvent method up to the incorporator stage except that it requires pure acetone as a solvent. It can be used for single, double and triple base propellants. The dough from the incorporator is extruded through a filter into convenient size cords and dried until the solvent is reduced. Warm rollers are then used to form the propellant into sheets prior to final extrusion after which stage the remaining solvent is removed by stoving. This method uses less solvent and results in a better product ballistically.

Solventless method. Used for double base propellants only, this method involves the use of heat and mechanical work for gelatinising the NC instead of a solvent. The paste so formed is dried and passed through heated rollers prior to extrusion, which results in greater precision in the final dimensions than the other two methods. There are, however, certain restrictions in the size of grains produced in this way.

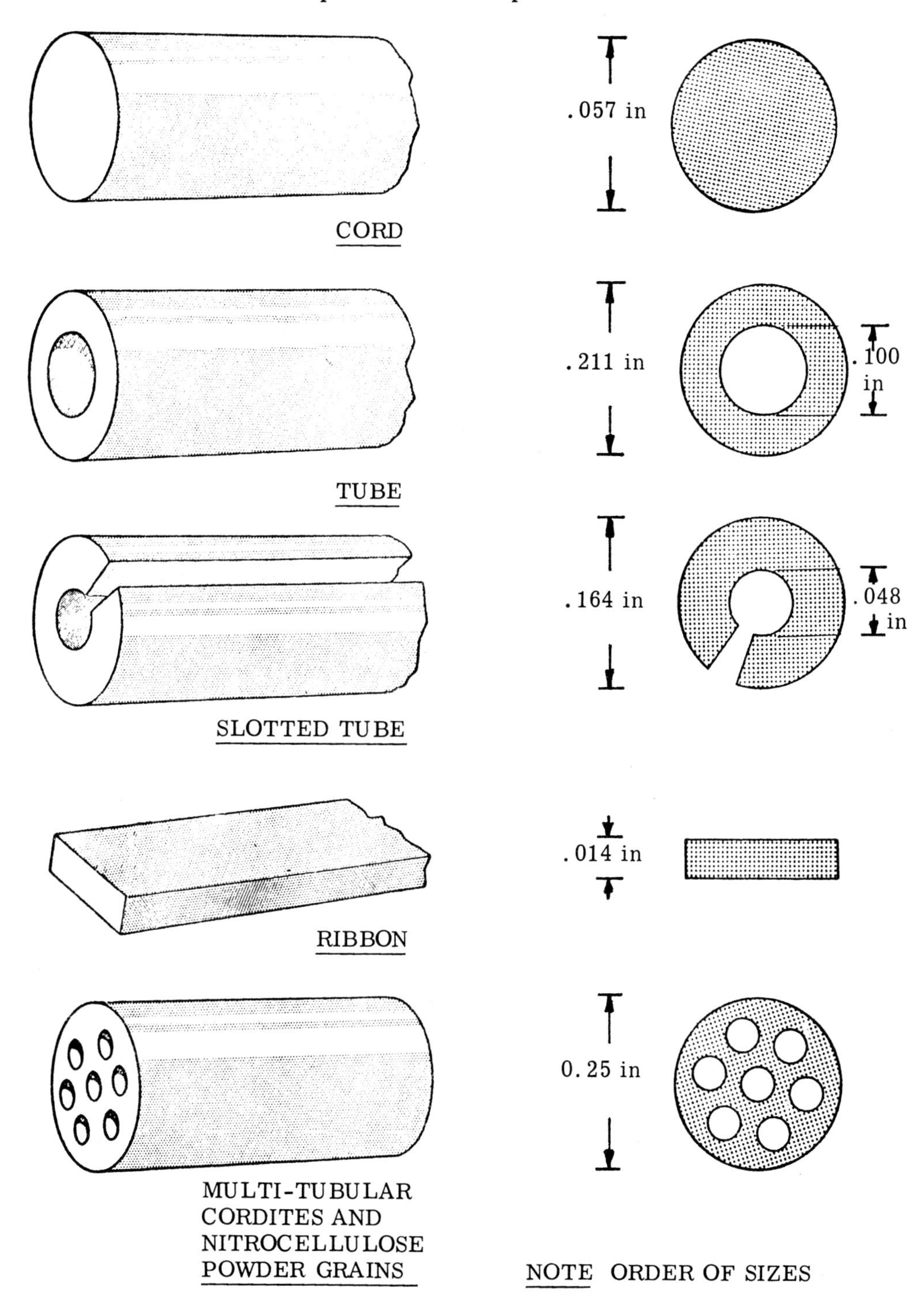

Fig. 7. Typical gun propellant shapes and sizes

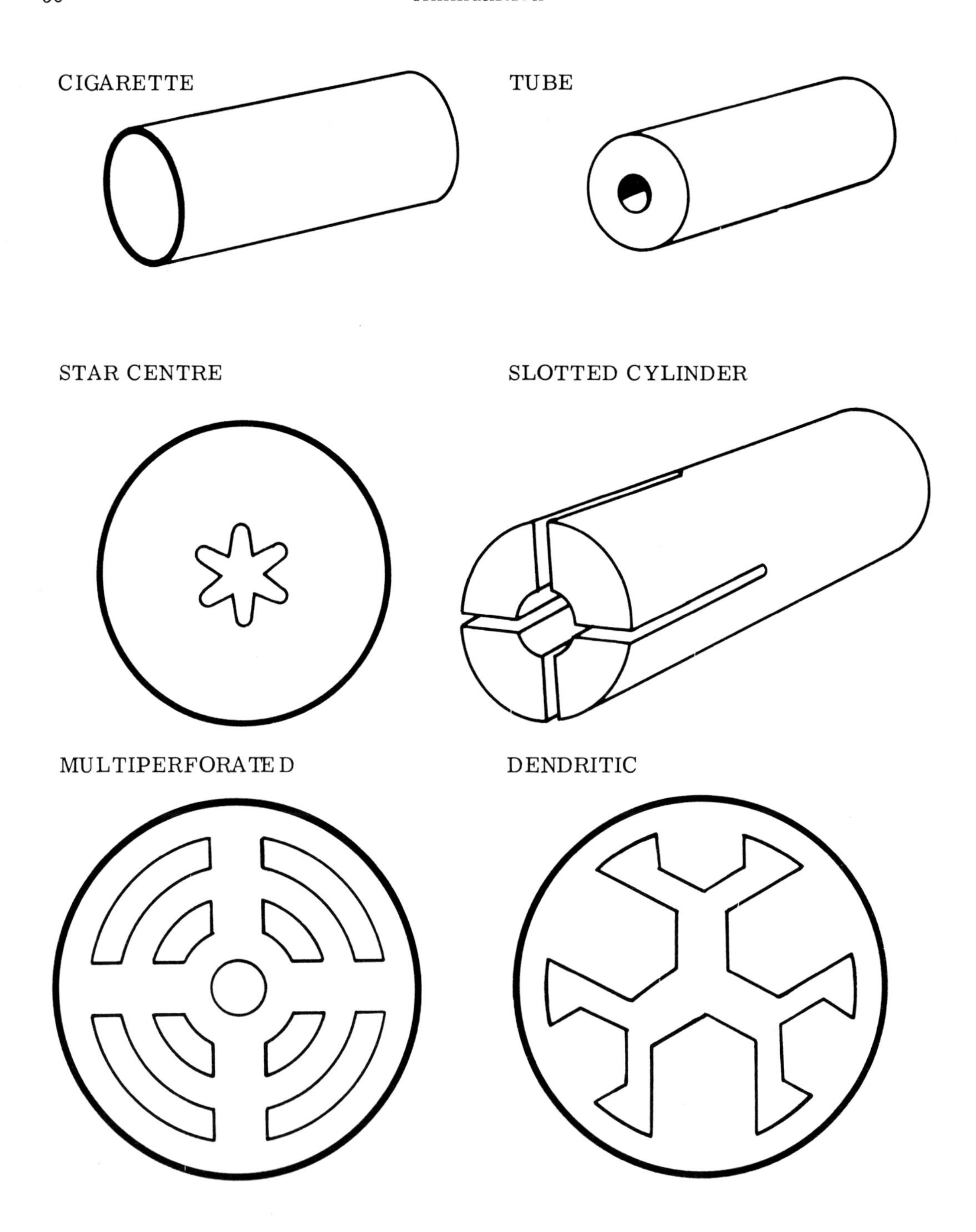

Fig. 8. Typical rocket propellant shapes

ROCKET PROPELLANTS

Solid Propellant

Chemical propellants provide the simplest and most efficient means of missile propulsion. They are relatively cheap, light and compact. There is no essential difference between these and gun propellant but they are designed to burn at much slower rates. Examples of sizes and shapes are shown in Fig. 8. Solid propellants are of two main types; colloidal and composite.

Colloidal solid propellant. Are essentially homogenous, gelatinous propellants of the cordite type consisting of nitrocellulose and nitroglycerine. They are produced in extruded form for small motors and cast form for large motors.

Composite propellant. Comprises a fuel which provides a solid rubber-like matrix in which a finely divided solid oxidiser (ammonium perchlorate) is suspended. There are two main types: extruded (or plastic) and cast (elastic or rubbery).

Inhibition and Case Bonding

Solid propellants which cannot be bonded to the rocket motor case must be inhibited on the outere surface to prevent burning. Cellulose plastics provide useful sheaths or coatings for this purpose and are shown as thick black lines in Fig. 8.

Platonisation. To achieve stable and uniform burning of solid propellants certain lead salts are added which produce a characteristic plateau in the pressure time curve Fig. 9. This is known as platonisation.

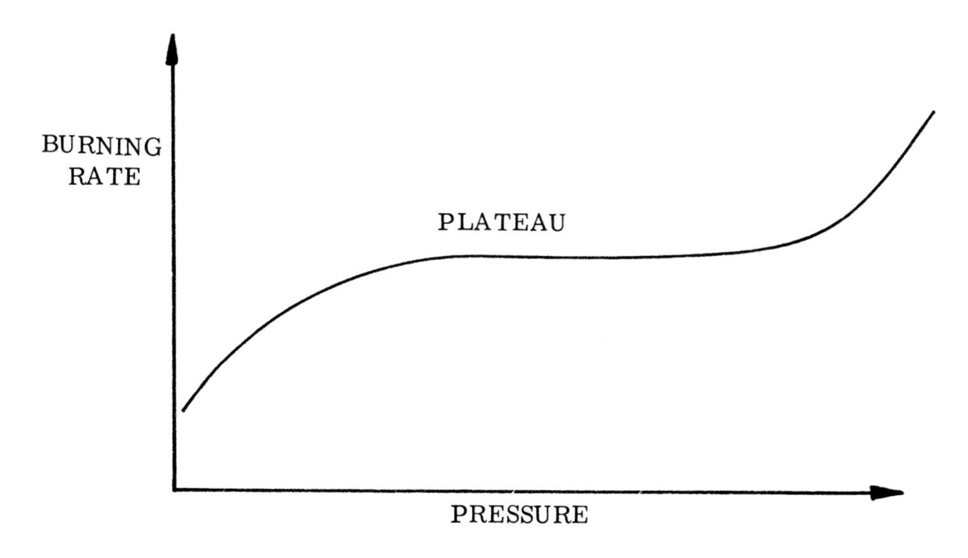

Fig. 9 Platonisation curve

Liquid Propellant

Liquid propellants are cheap, controllable and provide a high performance effect, but they tend to be more hazardous than solid propellants and require additional equipment for mixing and metering. There are two types: mono - which are combustible liquids burning in the absence of external oxygen of a comparatively low energy and used in missile power supplies; bi - consisting of two components, liquid fuel and liquid oxidiser which are injected separately into the combustion chamber and are normally used in large rocket motors designed to take heavy payloads (space and satellite vehicles). Examples of mono propellants are hydrogen-peroxide and isopropyl nitrate. Bi-propellants include hydrogen or kerosine as fuels with oxygen or fluorine as oxidisers.

Rate of burning - Rocket Propellants

The rate at which the propellant is consumed is an important parameter which determines thrust, acceleration and burning time. Solid propellants burn 'in situ' whilst liquid types depend on the rate of flow of propellant into the combustion chamber.

Fuel Air Explosives (FAX or FAE)

Fuel air explosives comprise volatile hydrocarbons which do not require oxygen for spontaneous combustion, continue to burn without oxygen or air, contain a high proportion of oxygen and cause a violent reaction on contact with combustible material, explode on contact with moist air at ambient temperature, react violently with oxygen-rich materials and can ignite spontaneously on contact with certain substances. These have a very much enhanced effect over the traditional high explosives and can be several times more effective. In proximity to the target a container of the liquid is ruptured by an explosive device, forming an aerosol-vapour cloud. The cloud is then detonated by a delayed-action igniter with delays from 100 milliseconds to 4 seconds depending on the type of weapon. This type of explosive is essentially an area weapon with blast as its main terminal effect. Detonation velocities and brisance are inherently low.

SELF TEST QUESTIONS

Question 1 What is gunpowder?

Answer ....

....

Question 2 What is the basic difference between a high explosive and a propellant?

Answer ....

Question 3 Explain an explosive train.

Answer ....

....

....

Question 4 What is an initiator?

Answer ....

Question 5 What are the two types of initiation?

Answer ....

Question 6 Describe the velocity of detonation of an explosive.

Answer ....

....

Question 7 Define the Figure of Insensitiveness of an explosive.

Answer ....

....

Question 8 What are the main ingredients of a single base, double base and triple base propellant?

Answer ....

....

....

Question 9 Why are propellant additives used and give examples?

Answer ....

....

....

....

....

....

Question 10 What is platonisation with regard to a propellant?

Answer ....

....

....

ANSWERS ON PAGE 249

5

Cartridge Systems

SCOPE

A cartridge system is part of a round of ammunition which provides the means of propelling the projectile to the target. Originally gunpowder charges were loaded into the muzzles of guns by scooping up loose powder in a long handled scoop and pouring it in. Later, in order to speed up the process and make it more consistent and safe, a measured quantity was filled into a cloth envelope, permeable to flash, and inserted. The quest for consistency in muzzle velocity and thus range is still of prime importance and this is the significant aspect of cartridge design. The main components of a cartridge system are the propellant charge which provides the energy, the primer or tube which ensures adequate initiation of the charge when fired and often an igniter which assists the initiation process. Before discussing cartridges in detail, a note on gun development is necessary to give the weapon cartridge relationship.

GUN DEVELOPMENT

Early guns were all smooth-bore muzzle loaded (ML) types where the propellant was first loaded followed by the projectile. Guns were then distinguished only by pattern but when in the 1850s Armstrong's breech loaded gun was introduced, the term BL was adopted to distinguish this pattern. The services then reverted to ML due to deficiencies in the breech sealing arrangements but adopted rifled barrels thus bringing the term rifled muzzle loader (RML) into use. In the late 1870s the modern screw breech made its appearance and these are referred to now as BL. The breech mechanism is generally heavy and opens rather like an oven door. The sealing of gases rearwards from the propellant charge when the gun is fired is known as obturation and this is achieved by the gun. BL guns are loaded with a bagged charge system as shown in Fig. 1.

The loading process of BL rounds was not too speedy as 3 components had to be loaded namely the projectile, charge and tube and the opening and closing of the breech by hand was not easy on the arm. This no doubt influenced the search for a more rapid method of loading and firing.

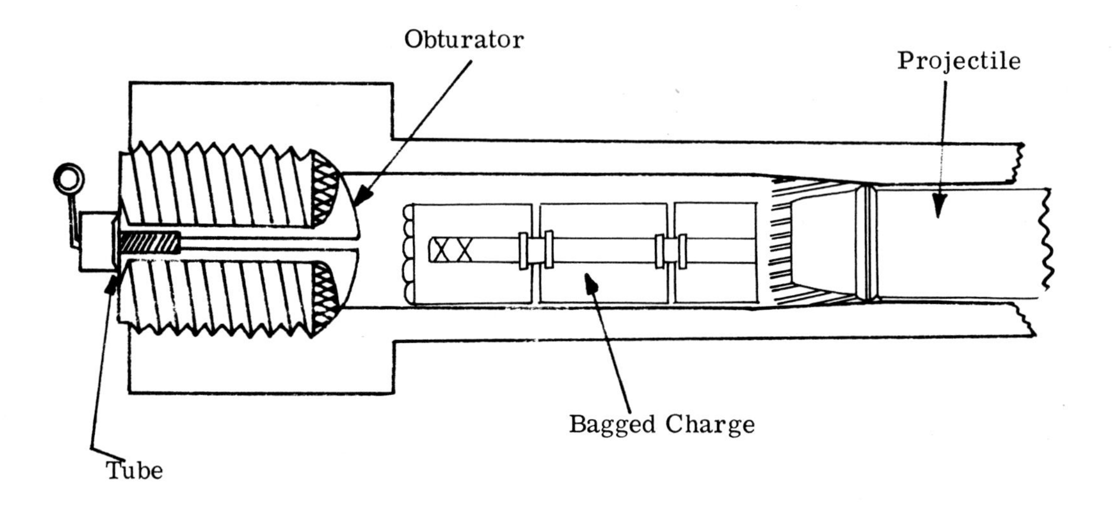

Fig. 1. Bagged charge

In 1881 came the advent of a light gun with a fixed non-recoiling mounting, a fixed brass-cased round of ammunition and a rapid breech of the sliding block pattern. This class of gun was named QF (for quick-firing), and an example is shown in Fig. 2. The obturation in this weapon is performed by the cartridge case.

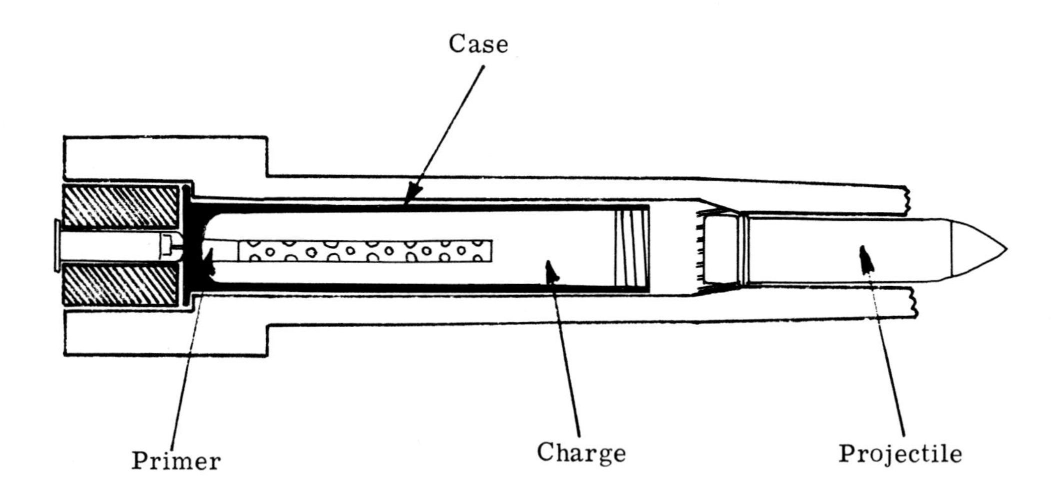

Fig. 2. Cased charge

The term QF probably owes its name partly to the simple action of opening the breech by a single sliding motion as opposed to the unlocking by rotation and swinging open the BL breech and partly because the round could be loaded in one piece. An added bonus came later where the breech was partly or completely opened automatically on run-out of the gun. Since then we have seen various gun/cartridge combinations and situations where BL guns can fire faster than QF guns. The abbreviations BL and QF have been retained to identify the obturation system in a gun and in recent years the terms "Bagged charge" and "Cased charge" have been adopted to replace the expressions BL and QF respectively when referring to cartridge systems.

Finally it was found necessary to develop a method of firing a heavy projectile from a relatively light weapon. This was due to the fact that BL and QF guns had to be sufficiently robust to contain, through obturation, the high pressures produced by the propellant gases when fired. The result was the introduction of a third system known as Recoilless (RCL). In this system the forward momentum of the projectile is carefully balanced by the rearward momentum of the propellant gases which are allowed to escape rearwards at a certain time thus obviating the need for a heavy recoiling mass. Cartridge systems for this type of weapon consist of a brass cartridge case fitted with a blow out disc at the rear to allow the gases to escape via a venturi in the weapon. An example is shown in Fig. 3.

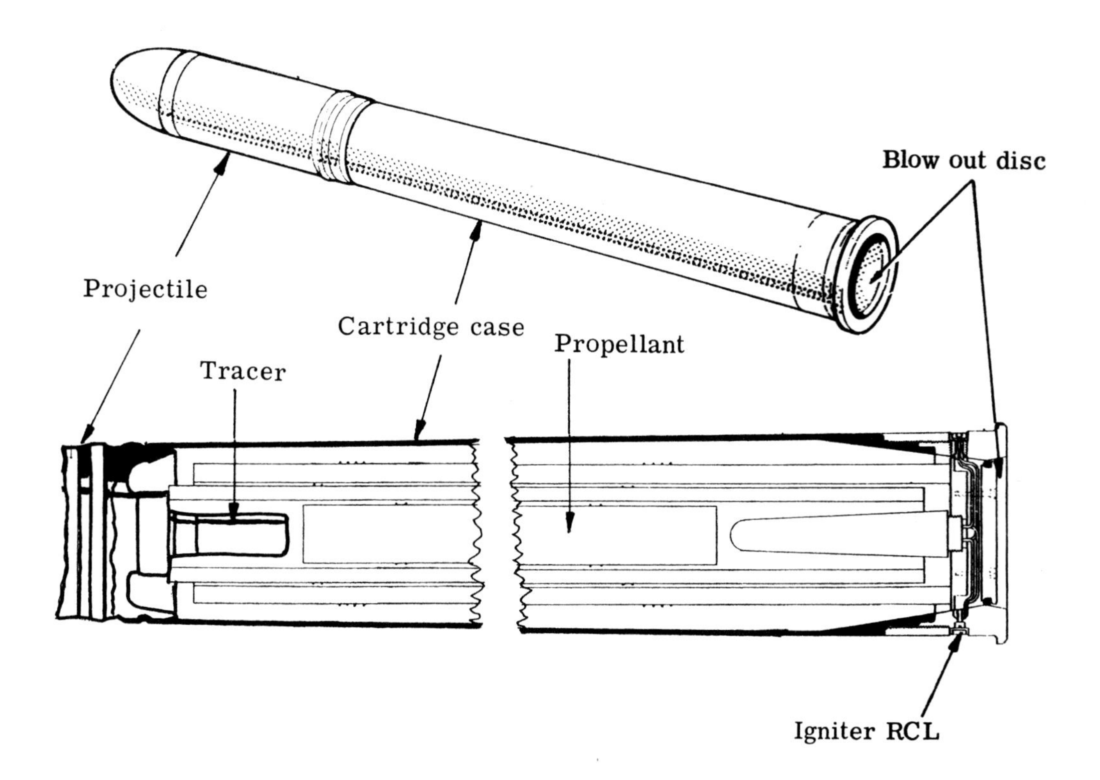

Fig. 3. RCL charge

CARTRIDGE SYSTEM DESIGN

Ignition

Explosive trains are covered elsewhere but it is necessary to refer to the importance of having an efficient and consistent ignition or starting point to achieve rapid combustion of the propellant charge. Systems can be either percussion or electrically fired although there has been one system which combines both. In either system the first explosive component in the train is a small cap containing a sensitive mixture which reacts to a blow from a striker or to energy from an electrical input. The spark, flame or heat produced, ignites the next explosive component, usually a magazine, and this boosts the initial effect thus ensuring adequate ignition of the main propellant charge. Some charges may have an igniter to assist rapid ignition of the main charge.

Bagged Charge System

In bag charges the propellant is tied in bundles which are then sewn into bags. Bag materials must be selected from those which do not leave residue in the weapon or smoulder when fired. They should also be compatible with propellant, resistant to rough usage, abrasion whilst in and out of the package, and totally consumable. Traditionally coarse silk has been used in this country but now certain synthetic materials (terylene/cotton) are also used. Terylene alone leaves a very hard glass-like residue which is not acceptable, but the addition of cotton tends to counteract this disadvantage and aids burning. Terylene, unlike cotton, is unaffected by propellant fumes and so retains its strength as a charge bag indefinitely.

Attached to one or both ends of the bag are igniters consisting of gunpowder or fine grain propellant fitted into shalloon 'compartments'. Shalloon is used because it is a woolly type cloth which retains the igniter composition yet allows the flash to pass through it on firing. More recently silk has been used for igniters fired in weapons using steel obturators where shalloon was found to be unsatisfactory. The materials used are not waterproof. Bagged charges may be filled with one or more types and sizes of propellant and have several configurations which allow the charge to be adjusted by the user. Some examples are shown overleaf in Fig. 4. The numbers shown refer to the charge for a particular firing range. Bagged charges for very large weapons are usually divided into fractional portions for ease of loading and handling.

The final component in a bagged charge system is the tube which is rather like a small arms round without the bullet. It can be initiated by percussion or electrical input and an example of each is shown in Fig. 5. Tubes are of the vent sealing type and are designed to seal the escape of propellant gases rearward through the vent. The flash from the tube should be as strong and as regular as possible to ensure the thorough ignition of the cartridge igniter. As the distance from the igniter to the tube varies with the type of charge in use, the strength of the flash must be sufficient to reach the maximum distance of travel required.

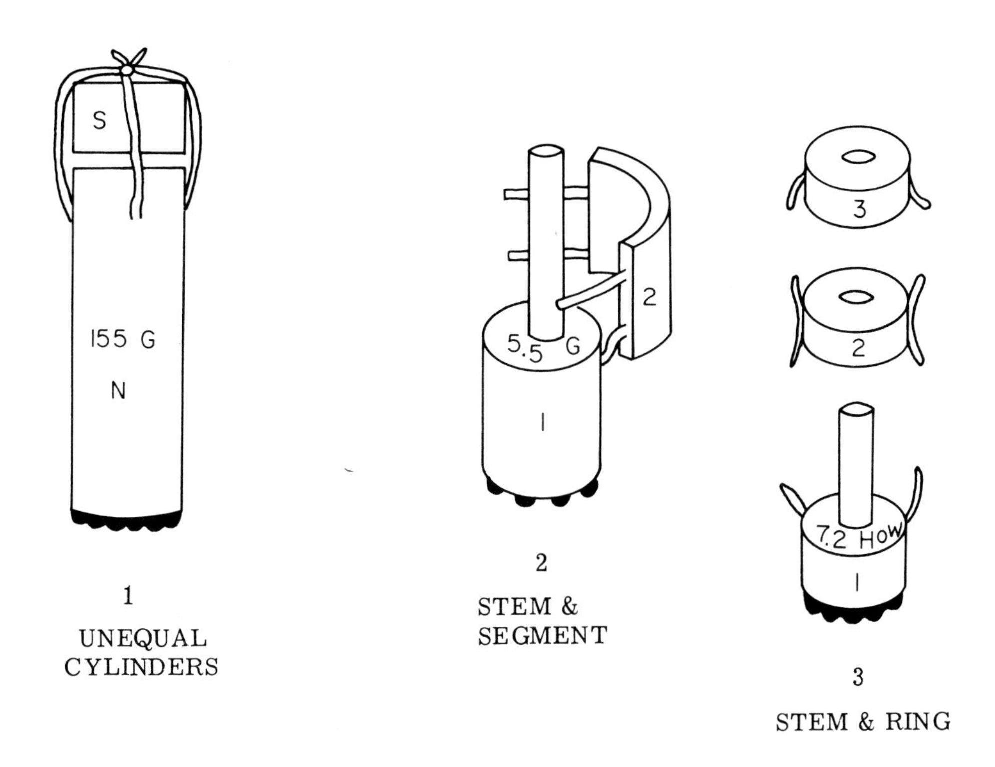

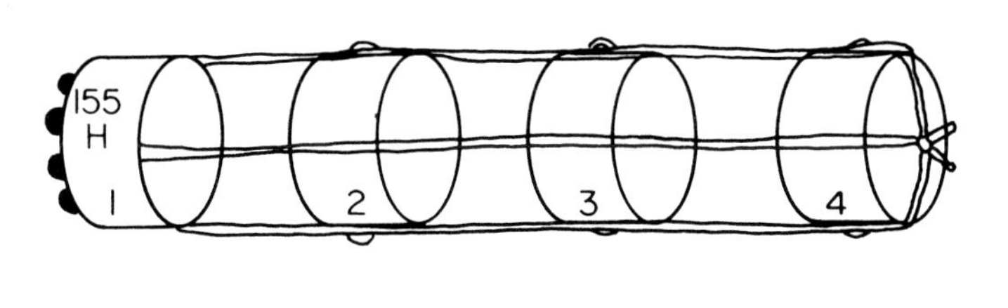

4

EQUAL CYLINDERS

Fig. 4. Bagged charges

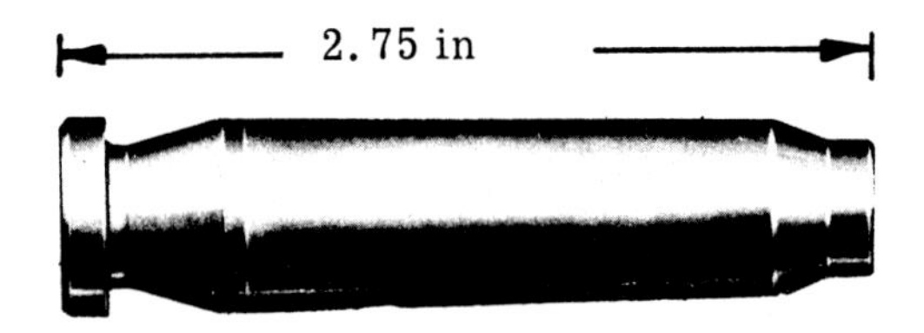

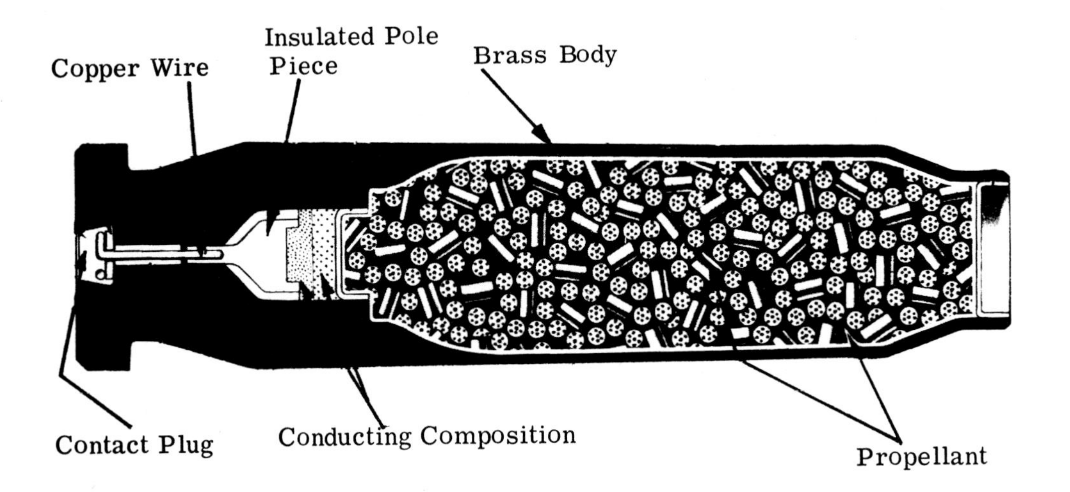

Fig. 5. Tube

Cased Charge System

In the cased charge system the propellant charge is contained in a tapered metal cartridge case which also provides the obturation by expanding under the heat of discharge thus sealing off all gas escape. An example is shown in Fig. 6.

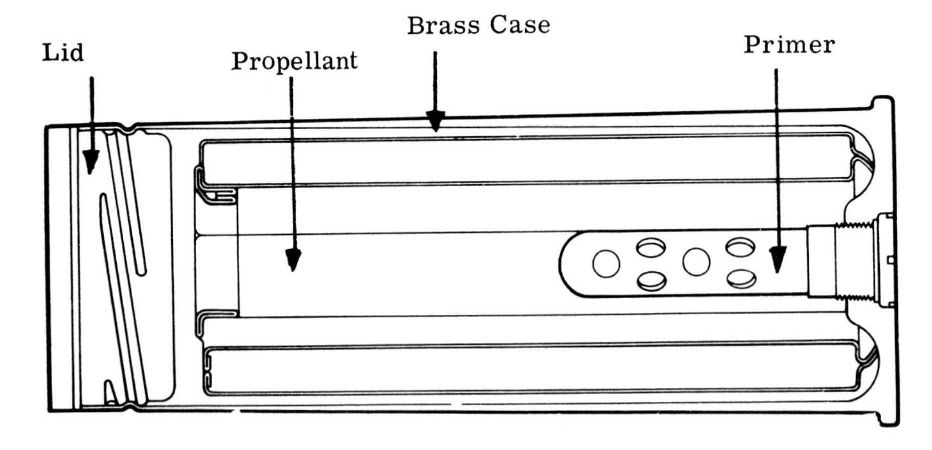

Fig. 6. Cased charge

The rear of the cartridge case is fitted with a primer which completes the obturation process rearwards and provides the initiation system. The primer can be either electrical or percussion and an example of each is shown in Fig. 7

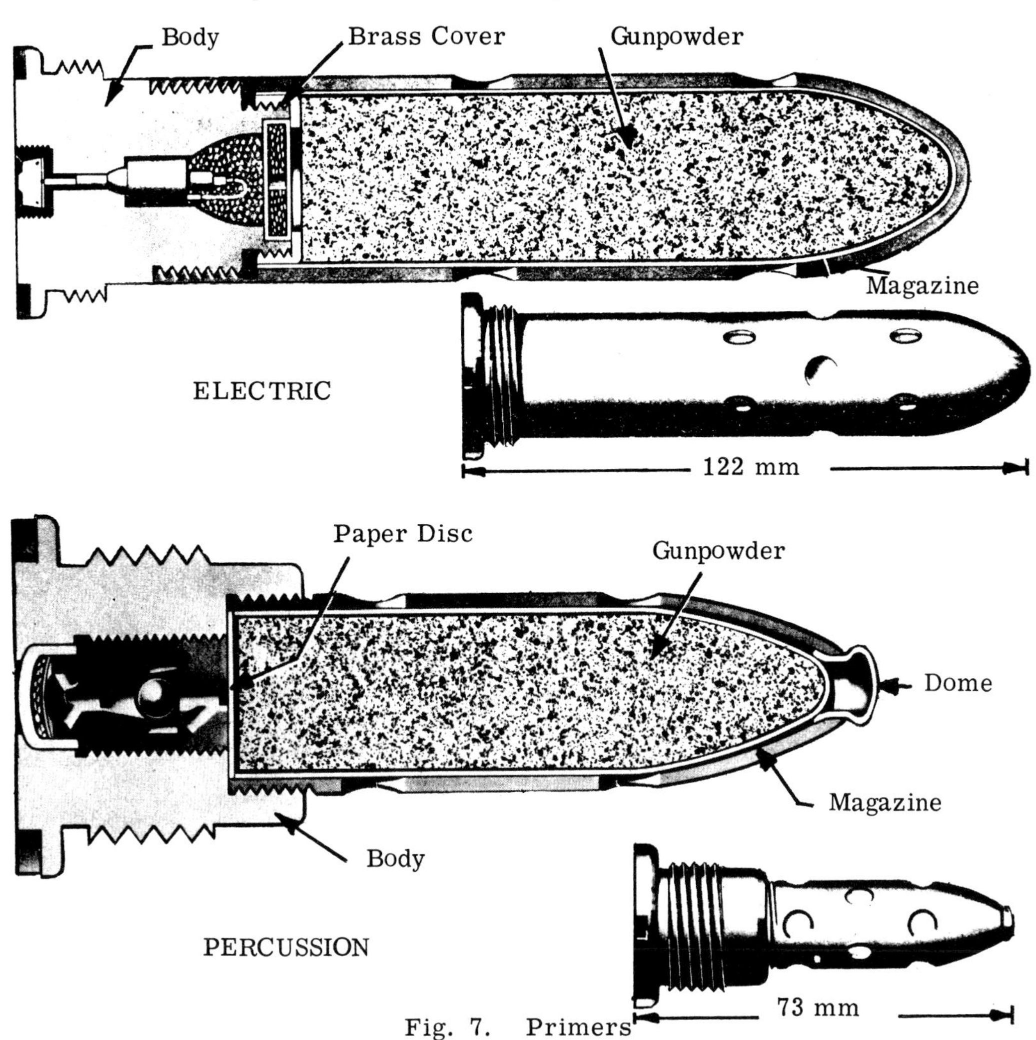

Fig. 7. Primers

Variations in Charge System

Although the bagged or cased charge systems are easily identified and well established in service there are more recent variations which have been added to each.

Recoilless Charges

Recoilless charges are classified as cased systems but do not obturate. The cases are fitted with igniters (instead of primers) and blow out discs although a combination of both a primer and igniter are found. Two examples are shown in Fig. 8.

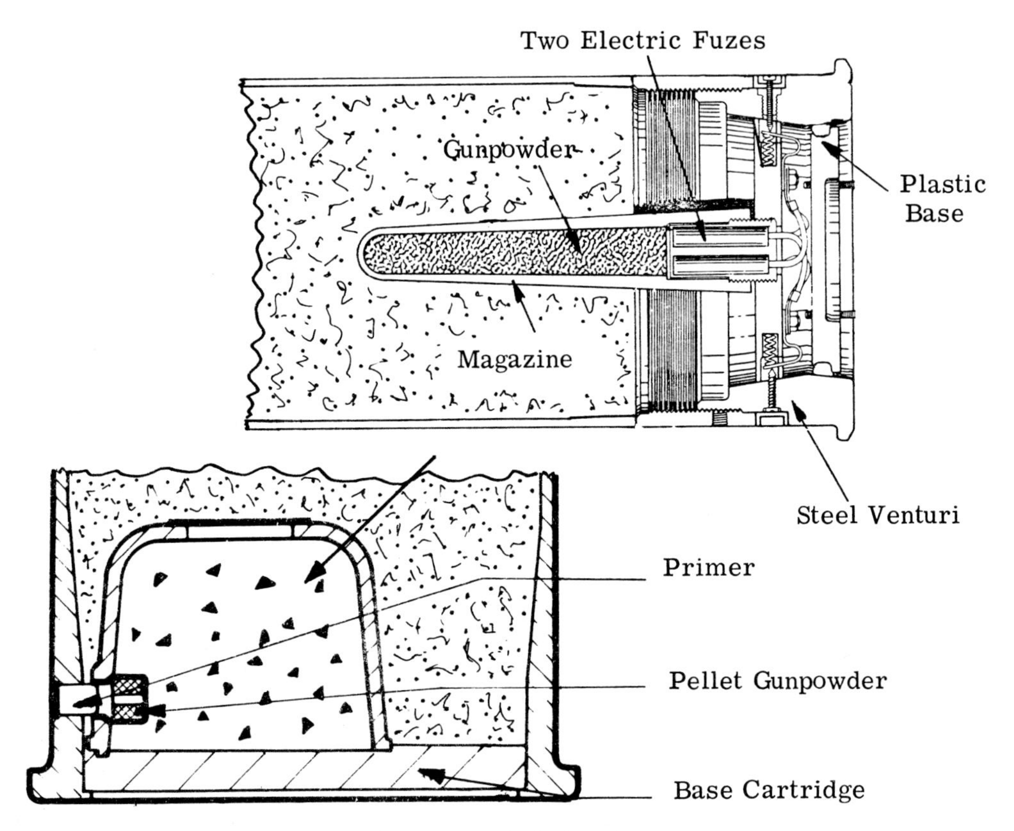

Fig. 8. RCL Igniters

Combustible Cartridge Cases

Combustible cartridge cases consisting of a mixture of kraft paper and nitrocellulose bonded with resin have replaced the cloth bag in some BL weapons. They are more rigid for ease of handling, provide better protection for their contents and are water resistant but are more easily ignited than bags. An example is shown in Fig. 9.

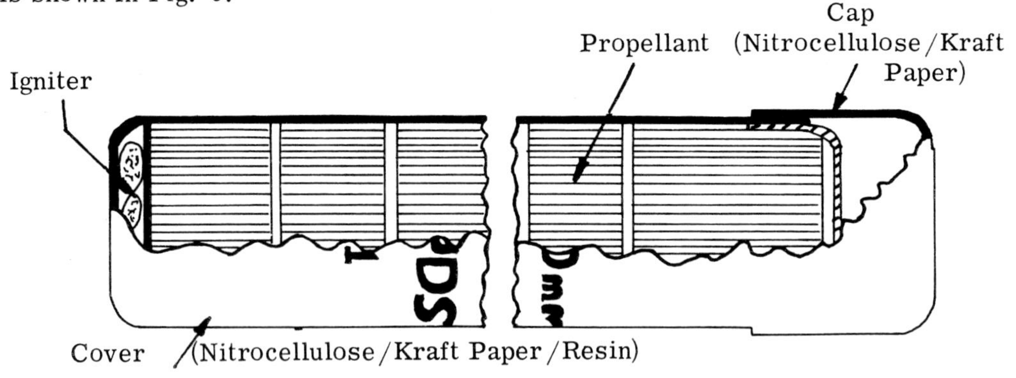

Fig. 9. Combustible cartridge case

Semi-Combustible Cases

Semi-combustible cases comprise a metal stub case in which is fitted a combustible case. The metal case also houses a primer thus linking it with a QF system. See Fig. 10.

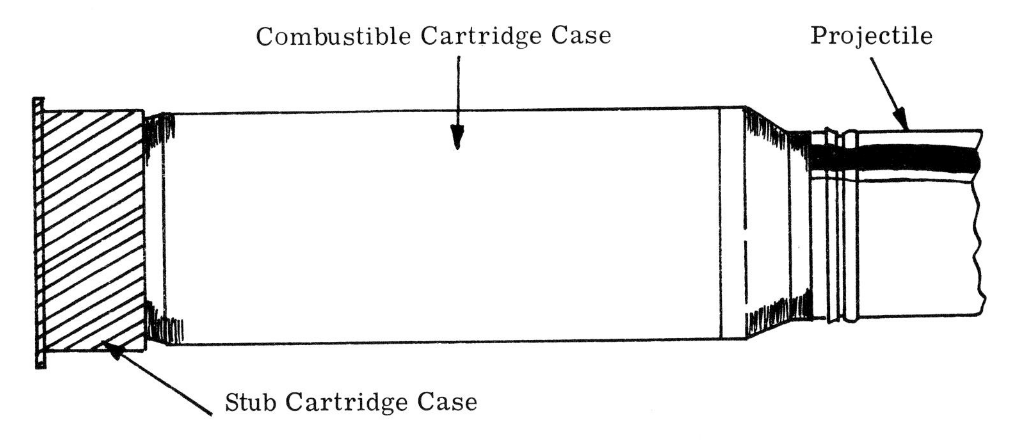

Fig. 10. Semi-combustible cartridge case

DESIGN OF COMPONENTS

Cartridge Cases

Cartridge cases are usually made of brass consisting of 70/30 Copper/Zinc, but other materials such as steel and aluminium have been used. The requirements of a case indicate the important part it plays in charge systems. Among thesee requirements are the protection of the contents, the provision of efficient obturation, easy loading and extraction, and the housing of the primer. The base of the case must be strong enough to withstand ramming and extraction whilst the mouth must have the facility to expand and contract easily yet have sufficient strength to support rigidly a projectile when fitted. There are various methods of attachment of projectile/case joins and these are shown in Fig. 11. The manufacturing process is complex and shown at Fig. 12.

Cartridge Bags

Some details of cartridge bags have already been given earlier in this chapter but further aspects are now included. The materials used must protect and contain the contents yet be totally consumable and allow the flash from ignition to pass easily through. They should be non smoulder, non stretch, non wear and need to be insect, vermin and rot proof. Any ties or stitching used must comply with the above requirements if residue is not to be left in the weapon.

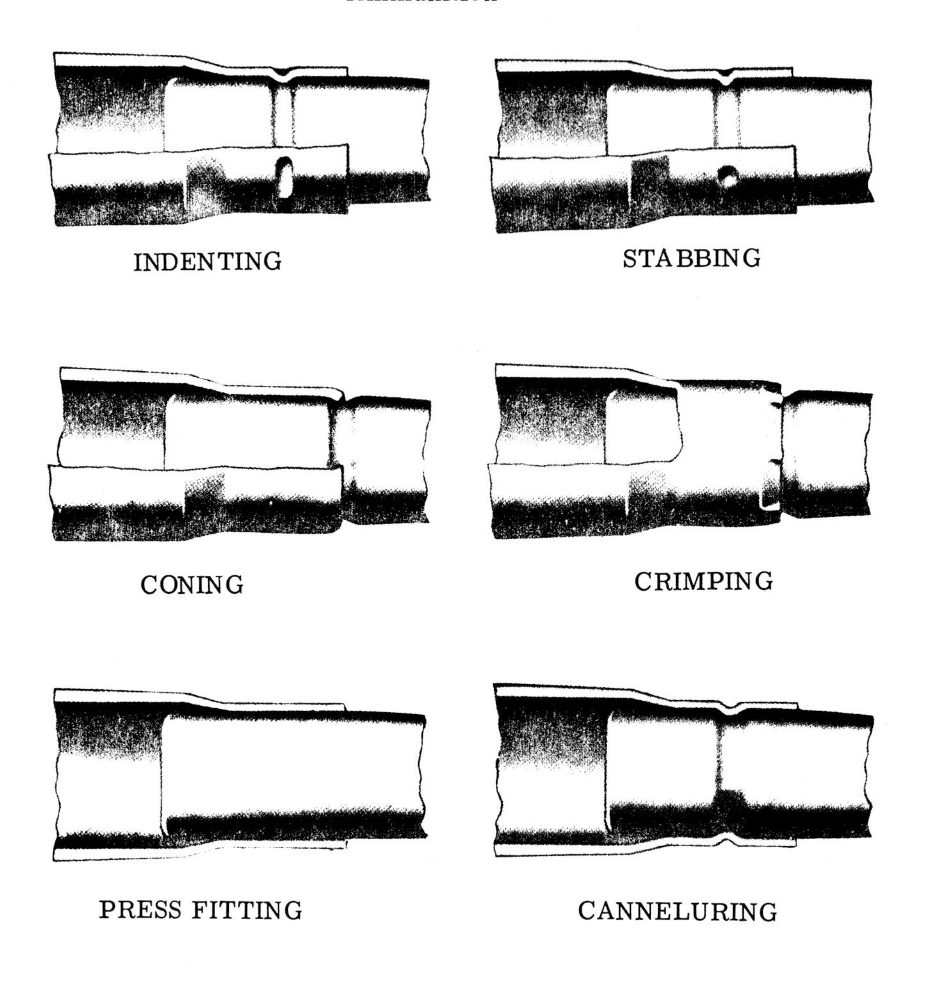

Fig. 11. Projectile/case joins

Igniter Bags

Igniter bags must be compatible with gunpowder or other propellant used and must allow an even spread of filling over the area of the parent charge. The material must not allow displacement of the igniter filling or migration into the cartridge or into the weapon when loaded.

Primers and Tubes

Primers and tubes are generally made of brass, the primer being either screwed or pressed into the cartridge case. The tube is merely loaded into the gun vent either singly or from a charger. The flash from the cap is often boosted by a powder pellet before igniting the main filling in the primer magazine or in the tube. The primer magazine has a paper or polythene envelope inserted to contain the charge yet allow the flash to pass through the magazine to the main propellant

Fig. 12. Manufacture of brass case

charge surrounding the magazine on ignition. Tubes do not have this arrangement as the effects of ignition must be uni-directional through the vent to the bag charge igniter. Percussion systems involve a 'one way valve' arrangement where the flash from the cap passes a pea ball or plug into the magazine yet cannot return when pressure builds up due to the forcing rearwards of the pea ball or plug, which blocks the channels. Figure 13 illustrates the systems.

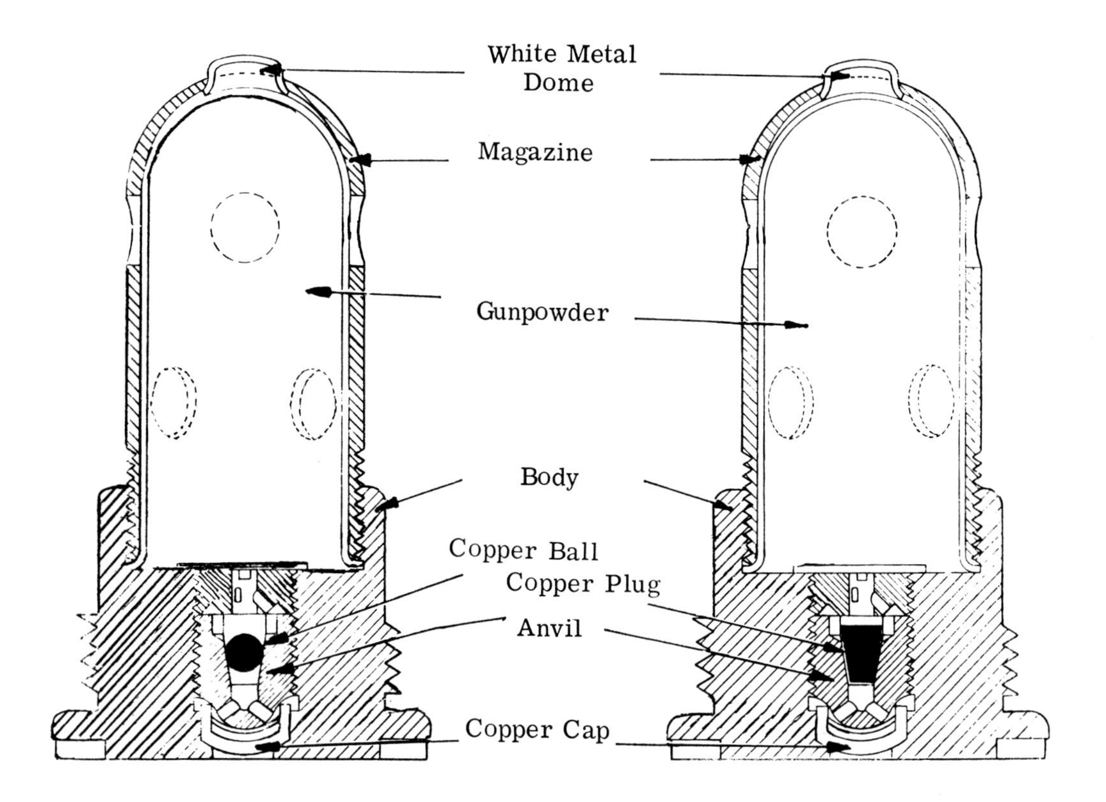

Fig. 13. Percussion primers

Electrical systems consisted originally of the heating effect of a thin wire surrounded by guncotton dust. This system was rather delicate and unreliable and current systems involve the use of conducting compositions in the cap which initiate when an electrical charge is applied. This system is more susceptible to electro-static hazards but can be screened or filtered to obviate such effects. An example of each is shown overleaf in Fig. 14.

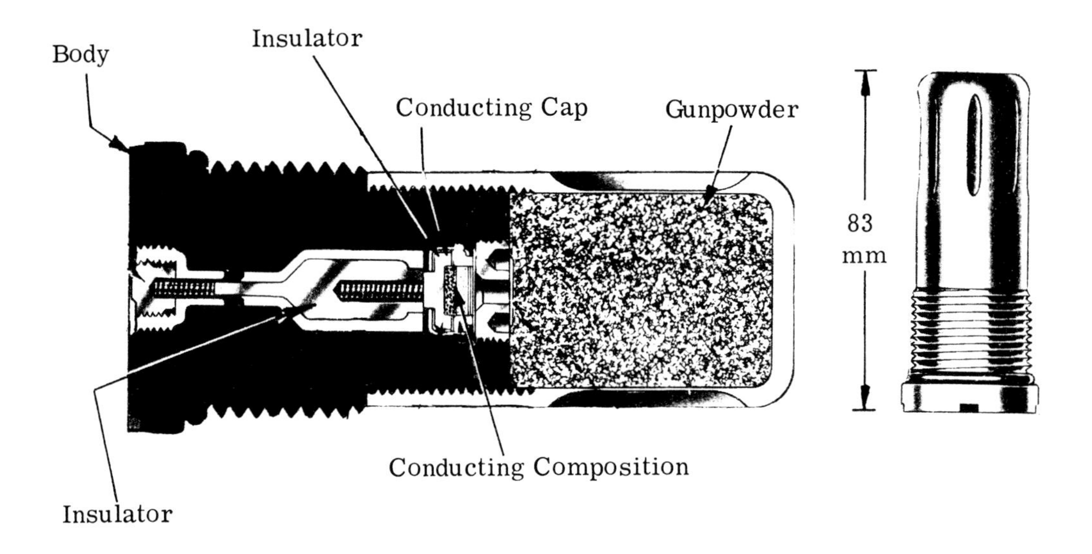

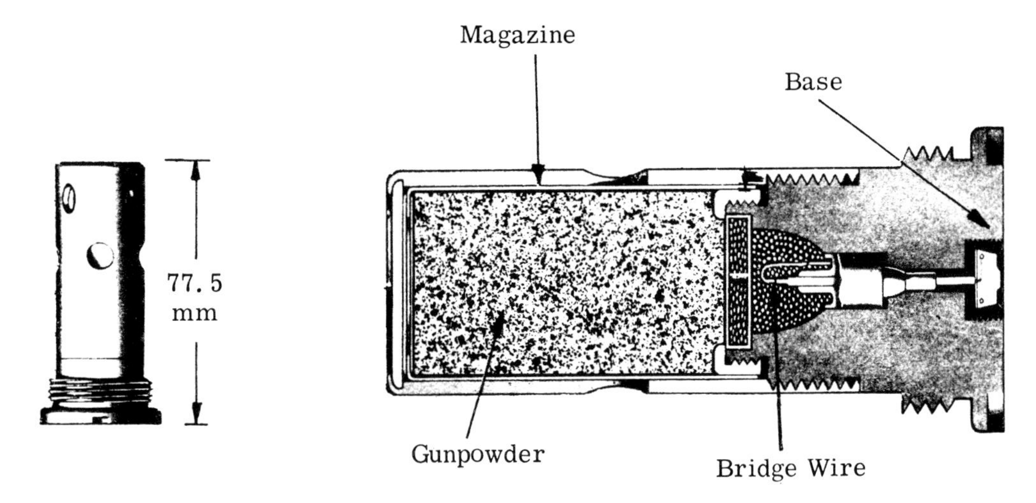

Fig. 14. Electric primers

WEAPON/AMMUNITION RELATIONSHIPS

The method of handling and loading cartridges into weapons influences the association of the cartridge to the complete round of ammunition. With large calibres it is necessary not only to load the projectile separately but often to load the charge in sections or increments. Some rounds if fixed as a complete round would be too long to handle and load. With weapons which fire at long and short ranges and high and low angles it is necessary to be able to adjust the propellant charge. On the other hand, quick firing and automatic weapons need a constant and rapid supply of ammunition. The weapon characteristics have resulted in various cartridge system arrangements. Bagged charges are easily adjustable and loadable and are designed accordingly. Cased charges are produced as fixed, semi-fixed or separate ammunition. Ammunition where the cartridge is permanently attached to the projectile is called fixed; where the cartridge is a loose fit to the projectile and packaged as such it is referred to as semi-fixed; and where the cartridge and projectile are packed and loaded separately it is known as separate. The semi-fixed round is not favoured in this country but is used in some other countries. The separate round allows the cased charge to be adjusted before loading into the weapon thus allowing for various range requirements. Some weapons use a super charge which can be a different charge entirely from the normal and some weapons use the same charge for different projectile types.

System Design Guidelines

Fixed rounds are likely to be chosen for automatic or direct fire systems whilst separate loaded cased charges are the more likely choice for small calibre indirect fire weapons. Large indirect fire weapons will invariably use a bagged charge system. Various factors must be considered at the design stage such as charge adjustment, rate of fire, handling, stowage, fumes, user environment and misfire drill.

Wear Additives

The effect of barrelwear is to reduce the life of the barrel and the muzzle velocity of the projectile. Relatively cool propellants are used in modern weapons but these do not completely solve the problem in high performance guns. A number of additives have been tried which either work on the principle of providing a cool gas layer between the propellant gases and the bore which reduces the heat input to the barrel, or reduce the temperature of the propellant burning. High density polyurethane foam liners stuck to the inside forward end of the cartridge case or wrapped around the forward end of a bagged charge have been successful. Fine talc with wax and wrapped in dacron cloth is another method and combustible cases have a similar effect. With small calibres an addition of titanium dioxide and talc to the propellant composition, although not as effective as the other method, is simple to use and does reduce wear to a certain extent.

CONCLUSION

There are several cartridge systems in use, all with their advantages and disadvantages, and no doubt more will be introduced in the future. The importance of the weapon ammunition relationship is well established and will influence any future designs.

SELF TEST QUESTIONS

QUESTION 1 Define a cartridge system.

Answer ....

....

QUESTION 2 What do you understand by QF and BL?

Answer ....

QUESTION 3 Explain the basic differences between QF and BL.

Answer ....

....

....

QUESTION 4 How does an RCL system operate?

Answer ....

....

....

QUESTION 5 What do you understand by ignition?

Answer ....

QUESTION 6 Describe a bagged charge and a cased charge.

Answer ....

....

....

QUESTION 7 How is obturation achieved in the charges at Question 6?

Answer ....

....

QUESTION 8 What is the difference in operation between a tube and a primer?

Answer ....

....

QUESTION 9 Explain the functions of a brass cartridge case.

Answer ....

....

QUESTION 10 What attempts are being made to reduce wear caused by propellant gases?

Answer ....

....

....

ANSWERS ON PAGE 250

6
High Explosive Projectiles

HISTORY

Early projectiles consisted merely of a solid iron ball propelled by a gunpowder charge. Many experiments were carried out in an effort to improve the destructive power of the projectile with little success until hollow projectiles were produced filled with explosive and initiated by a fuze. Early designs utilised an iron ball filled with gunpowder which was ignited on firing using a piece of slow burning match inserted in its side. This was a hazardous affair and gave very inconsistent results. With rifled barrels projectiles became cylindrical with rounded ogival heads but there were certain problems to be overcome. One of these was severe wear in barrels due to gas wash past the projectile.

Several designs of gas check were tried without success until a fixed gas check fitted to the base of the projectile and equipped with ribs or projections to fit the grooves in the barrel was introduced. This reduced wear and resulted in greater weapon accuracy. It was discovered that the gas check imparted spin to the projectile and this led to the driving band as we know it today.

REQUIREMENT OF A PROJECTILE

A high explosive (HE) projectile must be cheap, easy and safe to manufacture; it must be strong enough to withstand firing stresses, stable and accurate in flight and function efficiently at the target. Undoubtedly this raises a conflict for the designer.

At the Gun

When the gun is fired the projectile is accelerated along a relatively short barrel from zero to a few hundred metres per second in quick time and this results in a considerable set back force on the projectile. This dictates the thickness of the shell wall and base. The projectile is also spun in the same environment from zero to several thousand revolutions per minute and this involves pressure from the driving band on to the shell wall near the base plus rotational acceleration. With long projectiles, when fired in worn guns, there is a tendency to "buffeting" in the barrel which produces side-slap on the projectile. All these forces and others influenced the design of HE projectiles, see Fig. 1.

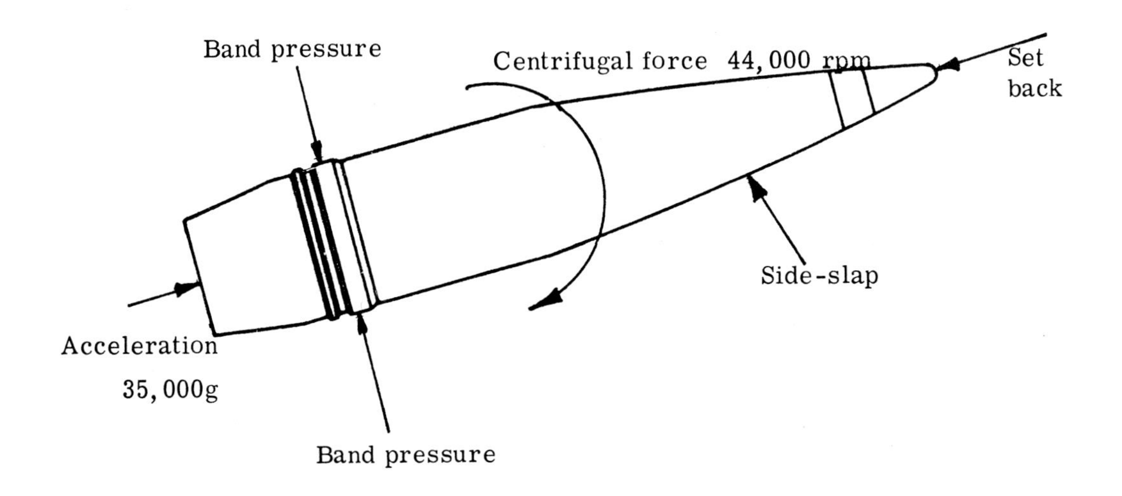

Fig. 1. Forces acting on a projectile

During Flight

Although this is mainly the problem of the ballistician, the behaviour of a projectile during flight is also the concern of the user from the accuracy aspect and does affect the quantity of HE carried and thus terminal effects. The main factors are stability, shape and carrying power. Stability of a projectile is achieved by spinning and provides accuracy with predictable range. Because the strength requirements at the gun inevitably give the projectile a base heavy effect it must be spun to ensure that the nose travels first. Stability ensures that the projectile will always point along the trajectory but there is a limit beyond which stability is not possible with current spin rates, if the shell is too long. Current spun projectiles are therefore restricted to about seven times the calibre length. Shape also plays a part in that a pointed nose carves more easily through the air than a blunt nose and a streamlined base has less drag than a cylindrical base. Where projectiles are fired with both sub- and supersonic velocities they are streamlined at nose and base. Effects of drag are shown in Fig. 2 where it will be seen that for subsonic velocities base shape is predominent but at supersonic values nose shape takes over.

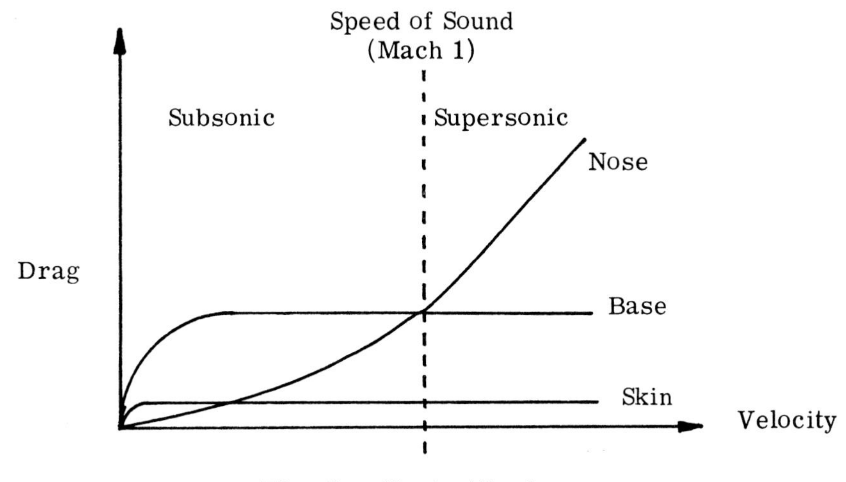

Fig. 2. Projectile drag

Carrying power is the ability of the projectile to reach the target and this depends quite naturally on its mass, shape and diameter.

At the Target

Having reached the target despite all the above constraints, the projectile must produce the required terminal effects as efficiently as possible. This involves good fuzing and initiating arrangements and an adequate filling of HE. There must be sufficient HE to break up the steel projectile into lethal fragments and project them consistently over a defined area. A typical HE projectile is shown in Fig. 3.

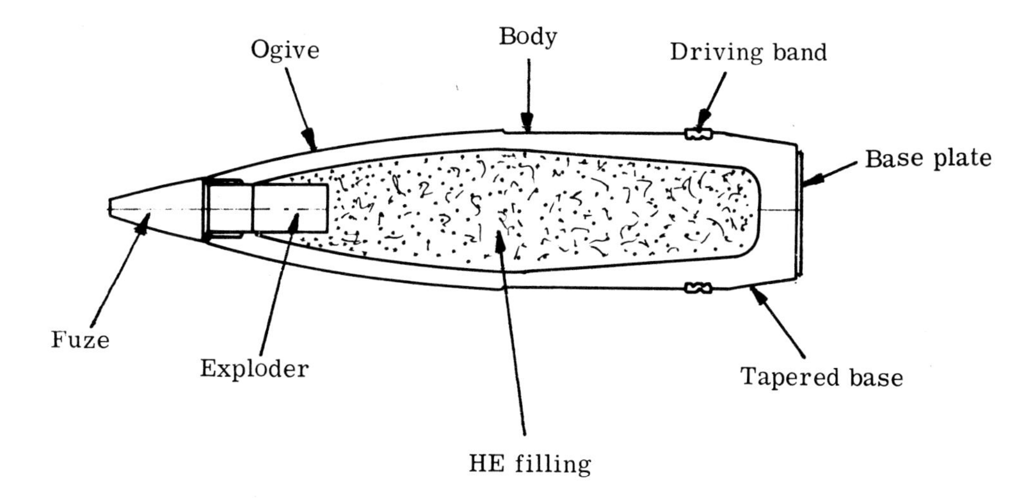

Fig. 3. High explosive projectile

REQUIREMENTS OF THE DRIVING BAND

The driving band is usually made of copper and positioned towards the rear of the projectile in such a way that it adds as little as possible to base drag and ideally should not project beyond the surface of the projectile body on leaving the gun. This ideal can only be met by discard at the muzzle and is only utilised in special projectiles. The driving band has several functions such as: to centre the projectile in the bore and rotate it; to prevent the forward escape of gases from the propellant charge; and to prevent slip back of the projectile when loaded at high elevations. The driving band also, through initial friction with the rifling, allows gas pressure to build up before rotating the shell. Examples are at Fig. 4.

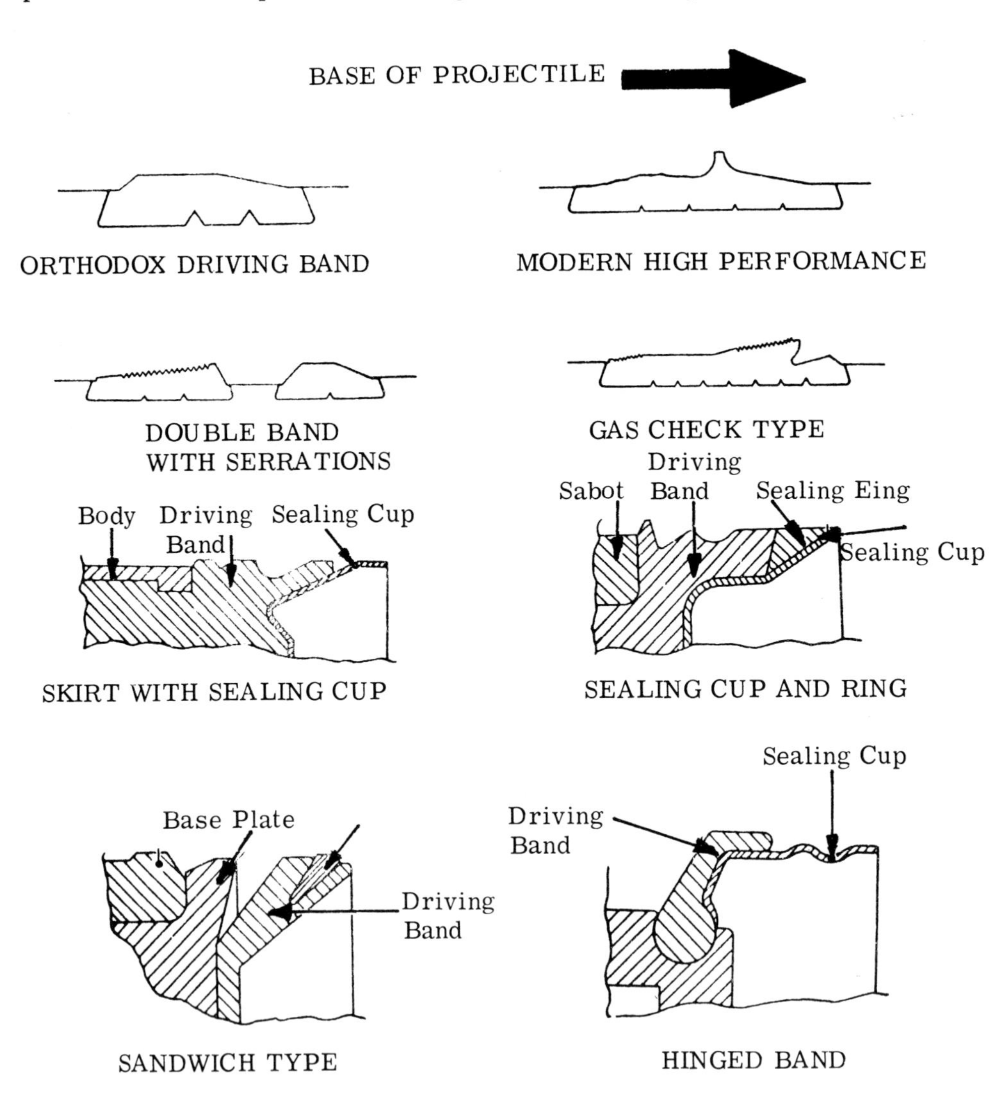

Fig. 4. Driving bands

REQUIREMENTS OF THE HE FILLING

The filling of the projectile is directly related to the cavity available and is referred to as the Charge to Weight Ratio (CWR). This ratio is the weight of HE over the total filled weight of the projectile and is shown as a percentage. Modern HE projectiles have a CWR of about 20-30%, this figure being set by the economic efficiency of the HE and strength of current steels. The type of filling used must stand up to the stresses at the gun yet be capable of efficient initiation at the target. It is referred to as the main filling or secondary explosive and sometimes the bursting charge. Choice of filling involves velocity of detonation, power, packing density and temperature requirements when being filled. The question of cost must also be taken into account particularly as the explosives with the best potential are usually the most expensive. Fillings are relatively insensitive and generally need to have a figure of insensitivity of a hundred plus.

Methods of Filling

These are strictly controlled and have been involved in several changes in the past. There are also likely to be further changes in the future. Filling designs include all explosive and inert components used in the projectile and ensure that projectiles are made and filled strictly to specification. Figure 5 from left to right shows the development of exploder systems which, in conjunction with the fuze, constitute the explosive train. Early designs had rubberised exploder bags which tended to reduce the explosive train effect due to the quantity of inert material involved in the bag closure (choke). Bakelite smoke boxes were also provided to assist the observer in locating the point of burst of the projectile and these became defective after long storage periods. This design gave way to the system of TNT/aluminium pellets and waxed paper wrapped exploder pellets. Stepped exploders were later used and it was thought that these might fracture and create problems. Modern projectiles have a universal cavity arrangement which caters for different fuze intrusions so that the user can, if required, insert a proximity fuze (deep intrusion) by removing a canned pellet or exploder.

Inert Components

There is a variety of inert components used in projectile cavity designs such as felt discs to cushion components and take up cumulative tolerances, tracing cloth discs to prevent damage to the pellet when fitting plugs or fuzes, and many others. All these must be compatible with the explosives and with each other.

Sealing

The object of sealing in the explosive filling of HE projectiles is to avoid contamination, ingress of moisture, or deterioration of the filling by environmental conditions. Contamination can be caused by exudation of the filling at high temperatures or by dust due to rough handling, vibration etc, during service use. Contamination of the fuze cavity can be hazardous, therefore adequate sealing is necessary. Usually the fuze or exploder intrudes into the main filling, the

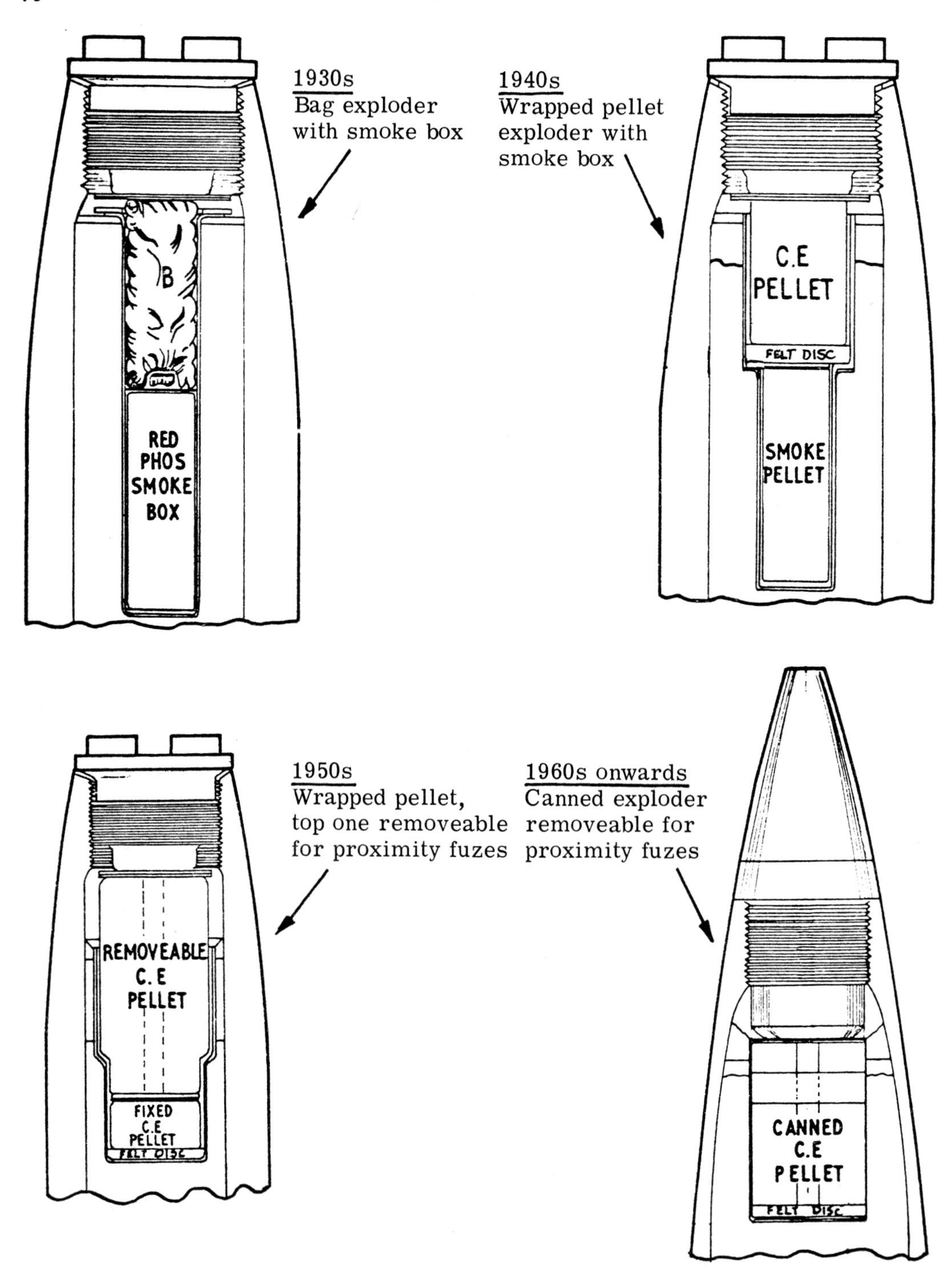

Fig. 5. Development of exploder systems

cavity of which is sealed with a paper liner or cup. The cup is varnished and sealed at the top by luting.

Filling Techniques

These consist of poured fill or pressed fill processes. Large quantities of compositions are accurately mixed from selected ingredients which have first been tested and approved. Each mix is identified as a 'lot'. During and after filling each is checked by extensive radiographic examinations. The explosive filling in a projectile should not be loose or cracked, neither should there be any cavities, piping or porosity, any of which could be hazardous. When the explosive is bonded to the projectile wall a difference in the expansion and contraction rates of explosive and projectile can produce defects in the filling, therefore great care is taken in the filling process to prevent this. One method is shown in Fig. 6.

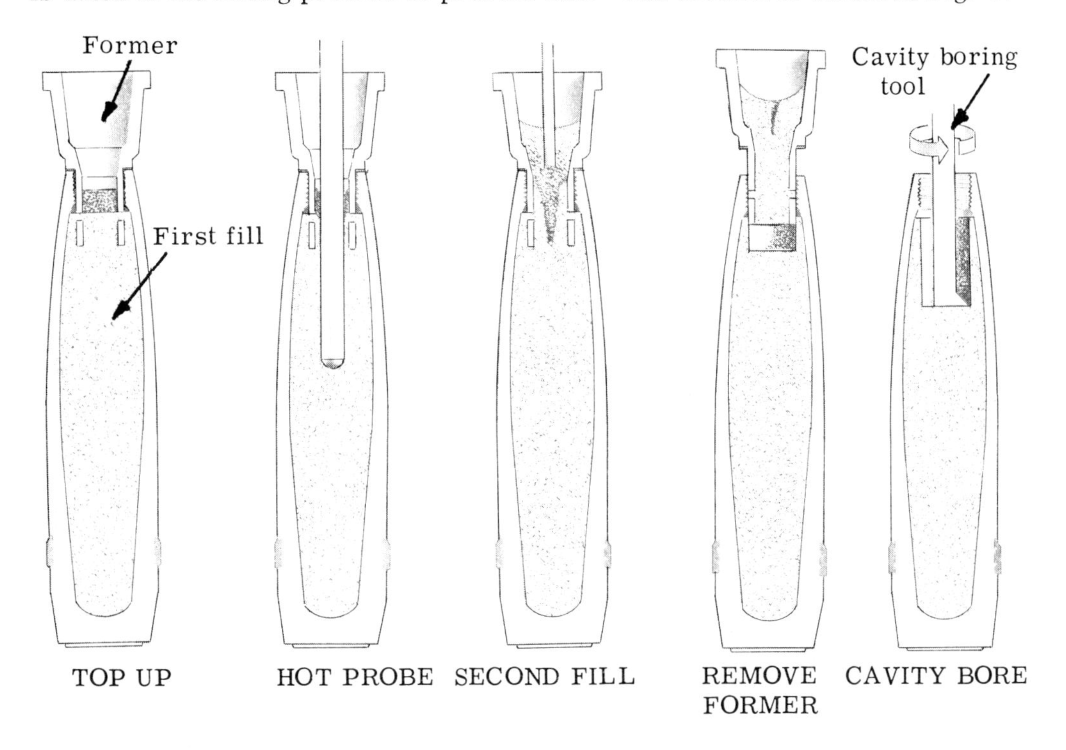

Fig. 6. Filling of typical HE projectile

Manufacture of Empty HE Projectiles

HE projectiles are usually made of steel because of its relatively low cost, adaptability, strength and toughness. They can be cast, forged or extruded. The casting process is unsuitable for modern HE projectiles as cast steel is prone to blow holes and centrifugal casting is at present a relatively expensive process. Forging is the main method used where steel is cast in a mould to form an ingot.

The ingot is then forged and rolled into bars termed bar stock which are cut into short lengths called billets. From each billet a shell is formed by forging. Induction heating of the billet keeps scale formation and other surface deteriorations to the minimum. A groove is cut into the wall into which the driving band is pressed and arrangements made for the attachment of the baseplate. The baseplate is needed to ensure that, in the event of any piping in the base of the projectile body, the hot propellant gases cannot penetrate into the HE filling when fired. Cold extrusion is a fairly recently developed process used for smaller calibres of projectile. All projectiles are identified and retain their identity throughout their life. The identity is linked directly to the bar or cast from which it was made. Figure 7 shows an example of forging.

FORGING

STAGE 1

Billet cut from steel bar

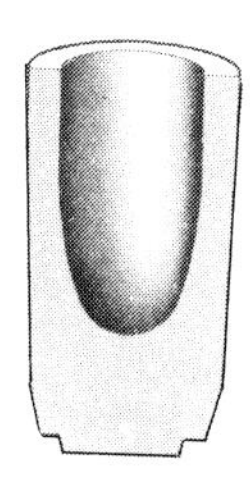

STAGE 2

After heating and piercing

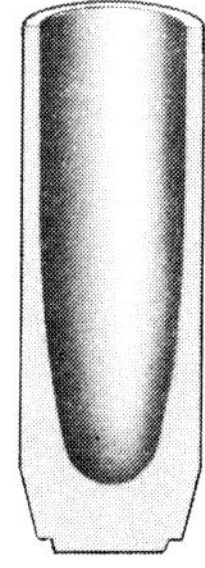

STAGE 3

After drawing

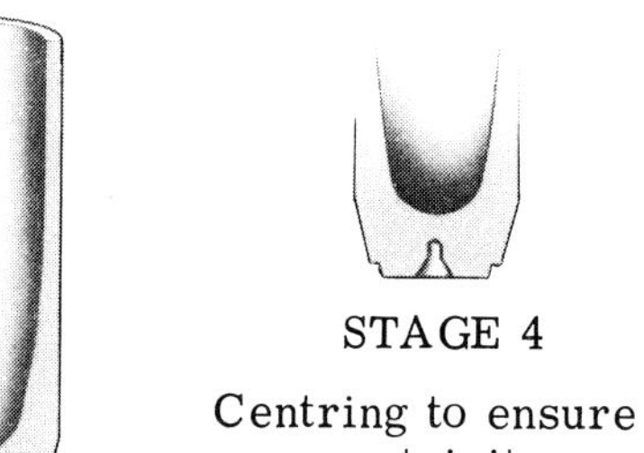

STAGE 4

Centring to ensure concentricity

ROUGH MACHINING AND HEADING

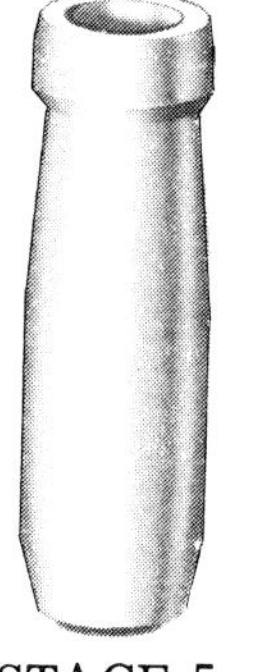

STAGE 5

Rough turning

STAGE 6

Heading. Formation of ogive in conical die after heating

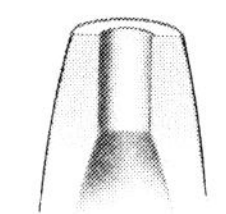

STAGE 7

Fuze hole roughly bored and finish turn body

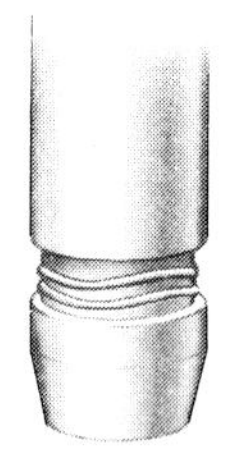

STAGE 8

Turn driving band groove

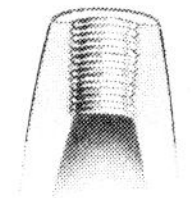

STAGE 9

Screw cut fuze hole

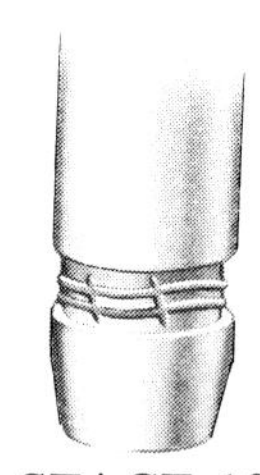

STAGE 10

Chisel cut waved ribs

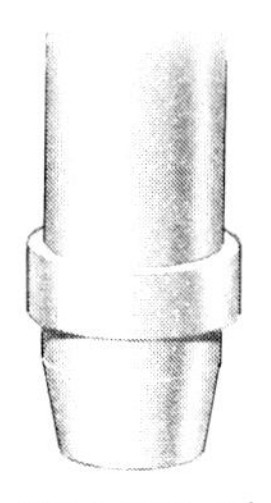

STAGE 11

Driving band pressed on

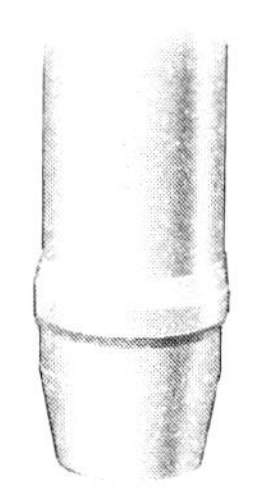

STAGE 12

Turn driving band

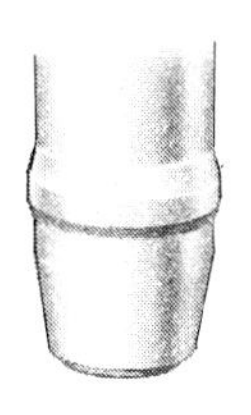

STAGE 13

Weld and face base plate

Fig. 7. Manufacture of HE projectile

CONCLUSION

HE projectiles have seen many changes in design but modern materials and filling techniques plus excellent quality control result in very efficient projectiles being available for service use. There will no doubt be further modifications and developments in the future which will ensure that the HE projectile, although designed to kill, will retain its very high standard of safety throughout the logistic chain.

SELF TEST QUESTIONS

QUESTION 1 State the requirements of a high explosive projectile.

Answer ....

....

....

QUESTION 2 Explain the forces and effects which act on a projectile

a. at the gun

b. during flight

c. at the target

Answer a. ....

b. ....

c. ....

QUESTION 3 What are the requirements of a driving band?

Answer ....

....

....

QUESTION 4 Define charge to weight ratio.

Answer ....

....

QUESTION 5 What is a method of filling design?

Answer ....

....

QUESTION 6 Why is it necessary to seal the HE filling in a projectile?

Answer ....

....

QUESTION 7 Name the two current filling techniques .

Answer ....

QUESTION 8 List the defects which should not be present in a HE filling.

Answer ....

....

QUESTION 9 Why is it necessary to have a baseplate fitted to a projectile?

Answer ....

QUESTION 10 Describe briefly the stages of forging a projectile.

Answer ....

....

....

ANSWERS ON PAGE 251

7

Warheads

INTRODUCTION

Warhead is now a generally accepted term and includes all types of projectiles such as high explosive, incendiary, nuclear, biological etc. "War" is perfectly reasonable but "head" is not, as many missiles have the main explosive charge in other than the head of the system. The object of the design and construction of any missile system is to deliver a warhead to the designated target or target area, the missile being merely the transporting vehicle. Although it can be argued that, if the missile actually hits a thin skin target it may not require a high explosive filling, the most accurate system is generally of limited value if the warhead is not correctly matched to the target. The warhead must produce sufficient lethal effect at the right time to destroy or incapacitate the target.

Early guided missiles adopted standard aircraft bombs as the explosive payload and these were encased in suitable frames and equipped with a form of guidance and propulsion. They were heavy and not initially matched to the optimum target requirement. Constant research and experimentation in warhead design enables designers to obtain optimum effects on targets of various natures. The warhead selected for a given target depends on the characteristics of the target such as type of armour, penetration required, speed, position, type of destruction required and so on. It is the air target which imposes most constraints on designers, although there are obvious problems for deep water target attack.

DESIRABLE CHARACTERISTICS OF WARHEADS

The desirable characteristics of warheads include efficiency, in view of the cost and complexity of the system, with maximum concentration of destructive elements on target area. Durability to stand up to the in-service environment is another, whilst simplicity, interchangeability and consistency are others. With regard to efficiency, the overall kill probability is the product of several factors such as probability of detecting target, correct launch, delivery to target and correct functioning. Lethality is the primary characteristic but it must be related to the conditions being considered.

TYPES OF WARHEAD

There are many different types of warhead and some will now be discussed.

Blast

Blast warheads have a quantity (usually large) of high explosive housed in a container. When initiated the detonation wave creates a high positive pressure followed by a smaller negative pressure. The wave moves radially outwards from the point of detonation at very high speed. Primary damage is inflicted by the pressure wave, accentuated by the negative effects and secondary damage results from flying debris. These warheads are of the internal or external blast types and are used against most types of target. Examples of these are shown at Fig. 1. The internal blast type is designed to hit or penetrate the target and when penetration of hard targets is required the warhead is designed with a toughened head so that it will penetrate and then detonate.

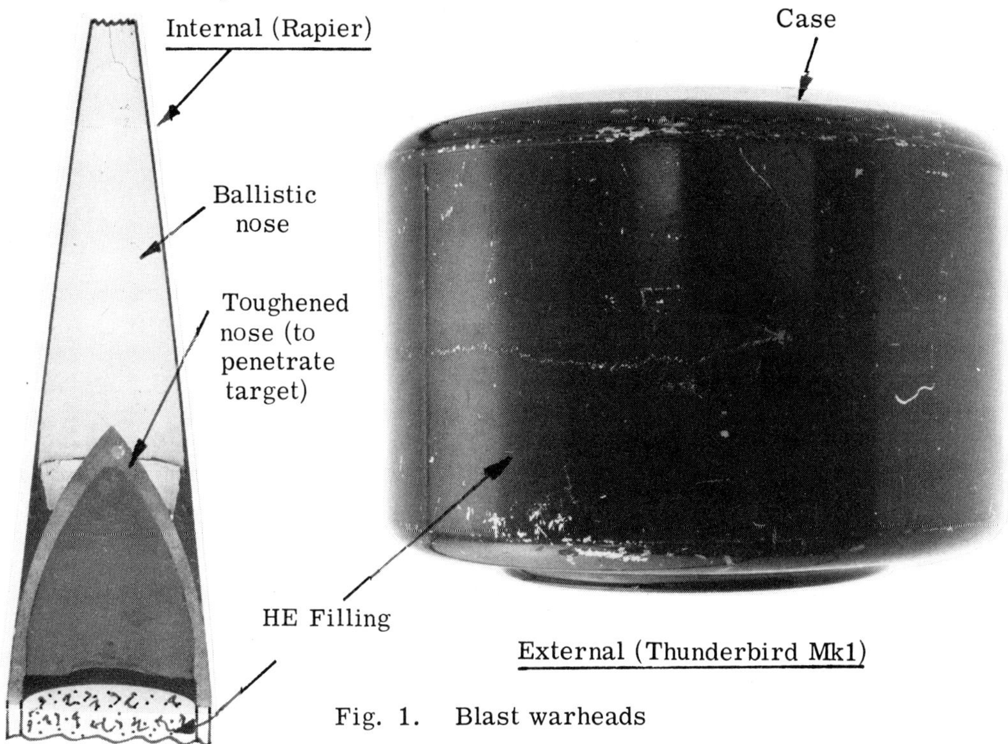

Fig. 1. Blast warheads

For soft skin targets the detonation pressure provides the penetration and damage effects. External blast warheads are designed to produce damage when detonated near the target and can thus be used in missiles with less accurate guidance systems. Proximity fuzes are used which are matched to the appropriate miss distance and lethal requirements of the system. Since the blast (pressure) effect of these warheads falls off rapidly with distance and altitude, they are best used at low altitudes preferably below 7000 metres and with miss distances of a few metres. On the other hand they are more effective against underwater targets as the greater density of water compared with air enhances the effect within limits.

Fragmenting

Fragmenting warheads are more closely related to conventional projectiles in that they rely on an explosive filling when detonation occurs to break up a metal case and propel metal fragments at high speed to the target. Due to the low launch acceleration for missiles compared to artillery projectiles the metal casing can be thinner and made up with metal fragments pre-formed to the required size and shape. Fragments tend to produce slow acting damage and although very effective against personnel, they are only successful against air targets if they damage the sensitive parts including the crew. Fragment weights may vary considerably depending on the required effects at the target and can range from a few grains to about 250 grains. The shapes can be cubes, rods, wires, spheres etc and are normally held together by a thin metal skinned container or resin. These are known as pre-formed fragment warheads. Fire-formed are those where the warhead is scored or notched on the inside so that it will break up in a particular configuration when detonated. Natural fragments are produced by a metal case without any notching where the detonation of the filling expands the case then breaks it up into random sized fragments. These warheads are more effective when detonated near the target rather than against it. Fragment patterns and effects depend on warhead profile and orientation to the target. Examples of different warheads are shown in Fig. 2.

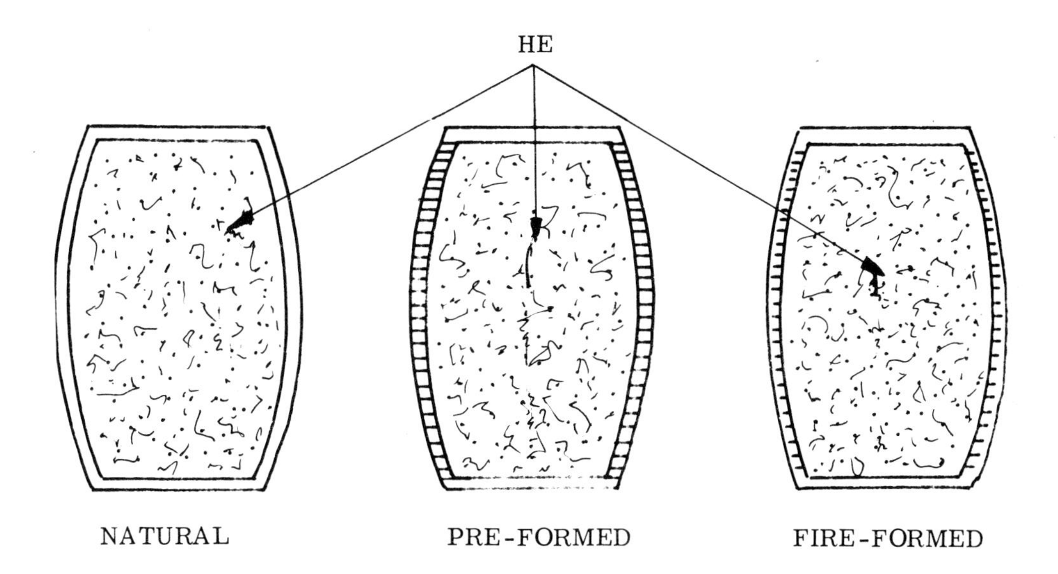

Fig. 2. Fragmenting warheads

Fragmenting/Blast

Most fragmenting warheads produce some blast effect at the target but the combination of both effects is an attempt to get the best of both worlds.

Discrete Rod

Small fragments are not the complete answer to air targets and, ideally, fragments which can slice through major assemblies or girder structures in air targets are likely to be more effective. Long rods surrounding a high explosive charge and propelled outwards on detonation of the charge would be most effective. The tendency to topple and lose velocity and to present arrow-like effects on the target make them little better than fragmenting warheads except at very high altitudes.

Continuous Rod

A development from the discrete rod by welding alternate rod ends together produced a very effective warhead, called continuous rod, for use against air targets. The rods are similar in length and thickness and are matched to the warhead size. The cross sectional profile may be circular, square or trapezoidal and the result of a complete hoop making contact with an air target is usually a rapid kill. The welding together of the rods is particularly important to ensure correct formation of the hoop and it is necessary to control the high explosive detonation speeds to the rods by wave shaping. An example of the operation of this type of warhead is shown at Fig. 3.

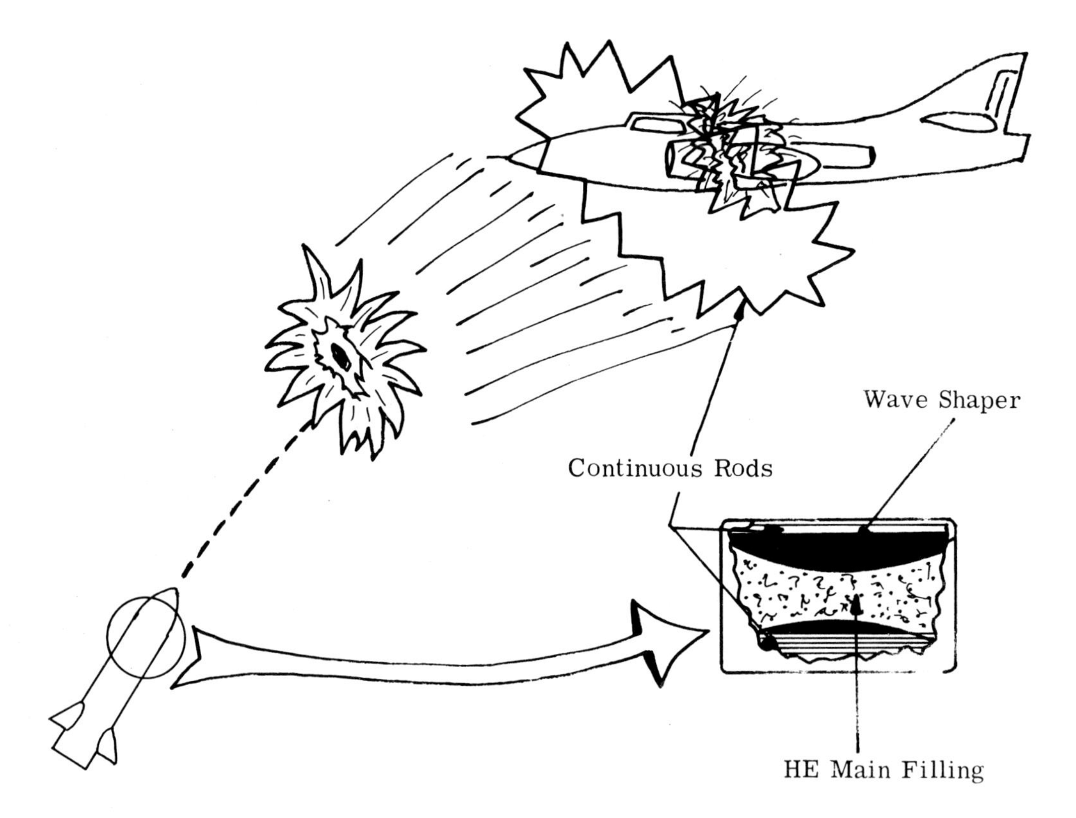

Fig. 3. Continuous rod warhead

Shaped Charge

The first of the directional warheads which consists of a thin case containing high explosive with a metal liner at one end. The liner is normally cone shaped but various other configurations have been used. The effect of this arrangement is to focus or concentrate a gaseous metallic jet of very high pressure at the target thereby penetrating it. It is mainly used against armoured targets but in modified form can be very successful against sea or air targets. The jet velocity can be of the order of 5-7000 m/s at the front end whilst the rear portion is slower and forms a slug having a velocity of 500 m/s.

Warheads can be fitted with one or more shaped charges and the orientation of such charges into a particular beam width with respect to the target is not easy to achieve.

Against armour targets the continuous jet produces penetration whereas with long stand off distances against aircraft the jet is broken and is thus able to produce wider damage effects. An example of a shaped charge warhead is shown in Fig. Fig. 4.

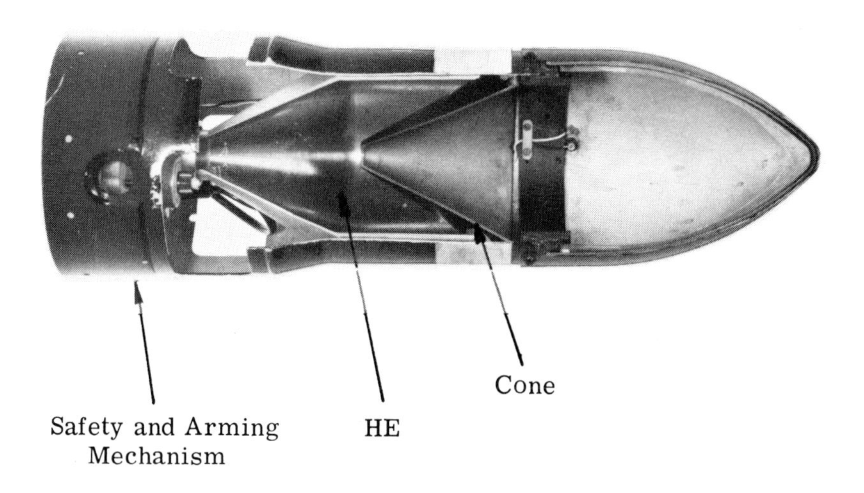

Fig. 4. Shaped charge warhead

High Explosive Squash Head

As the name high explosive squash head implies these relatively thin containers filled with high explosive and fitted with a fuze at the rear, squash on the target, usually armour, and the detonation wave passing through the armour produces a large scab on the inside due to the reversal of the wave at change of density. An example is shown overleaf in Fig. 5.

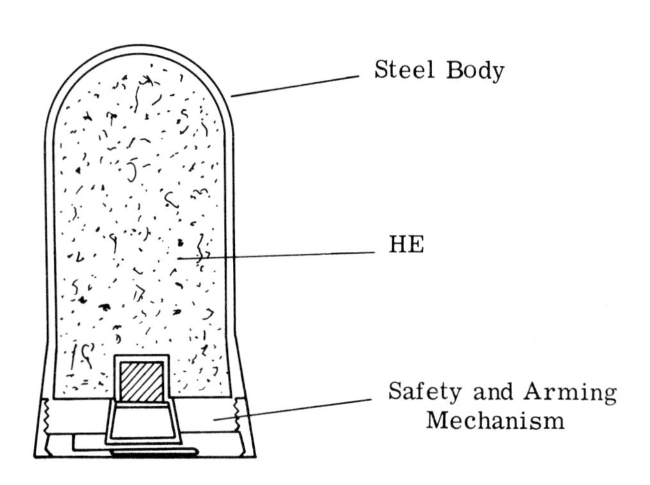

Fig. 5. HESH warhead

Bomblets

A bomblet warhead is filled with small containers or bomblets each capable of carrying its own initiation system. They are normally distributed randomly at the target when released from the parent warhead. Bomblets can be anti-personnel, anti-armour or they can be filled with various compositions solid or liquid depending on the requirements. The release of bomblets is usually attained by removing the outer skin of the warhead explosively thus allowing the bomblets to be thrown out by aerodynamic and gravitational forces. A typical arrangement is shown in Fig. 6.

Bomblets can be of various shapes and the shape selected will be influenced by the packing density required. They do not have their own means of propulsion but rely on their shape only to gain dispersion.

Sub Projectile

Sub projectile warheads contain a number of self-propelled sub projectiles. There are several shapes and sizes of such projectiles.

Cluster

Cluster warheads contain sub missiles which can be ejected from the warhead on individual ejection devices. Each sub missile contains an explosive charge and an initiating system. Used primarily against air targets, they can have from ten to several hundred sub missiles. An example is shown in Fig. 7.

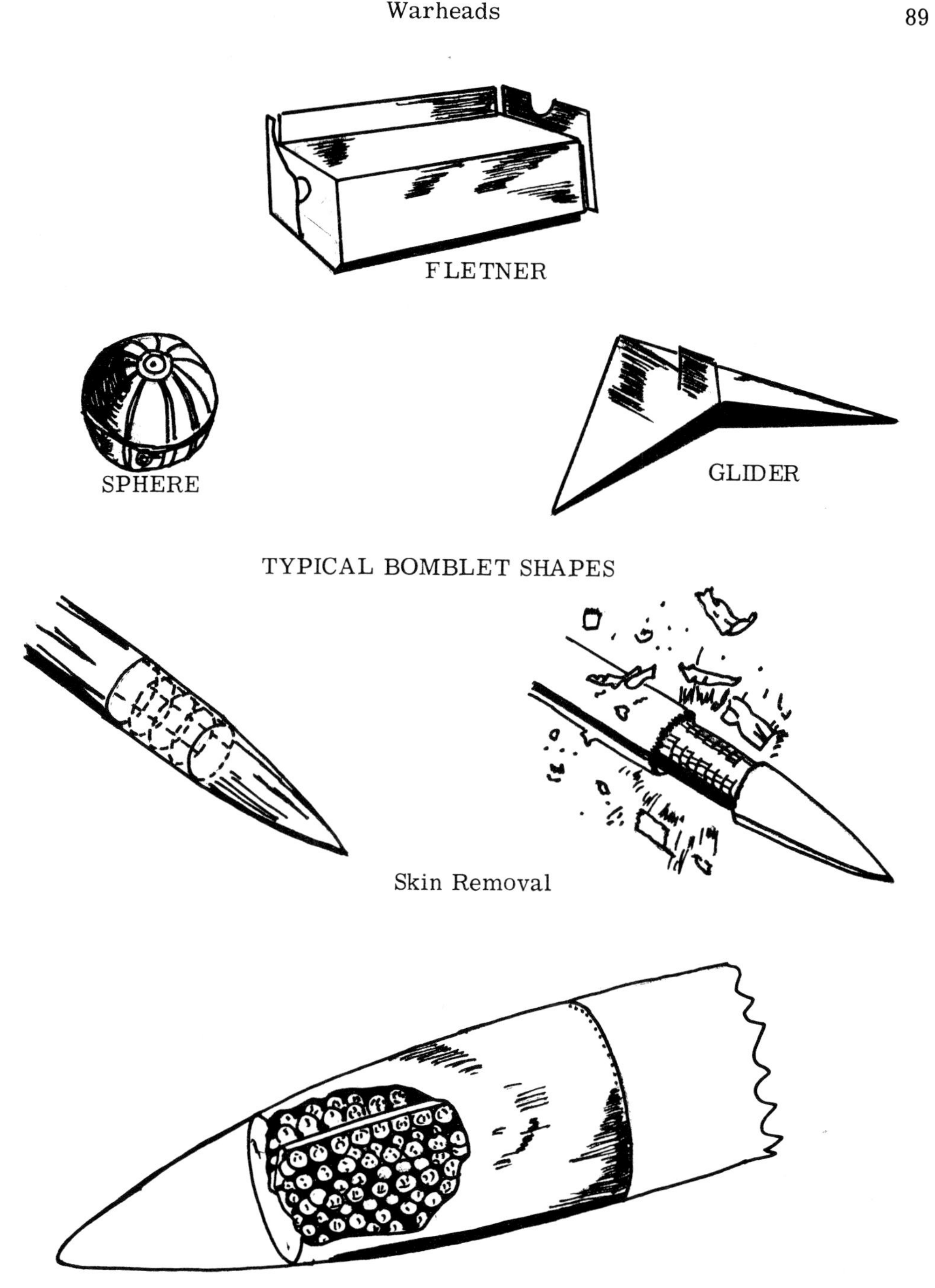

Fig. 6. Bomblet warhead

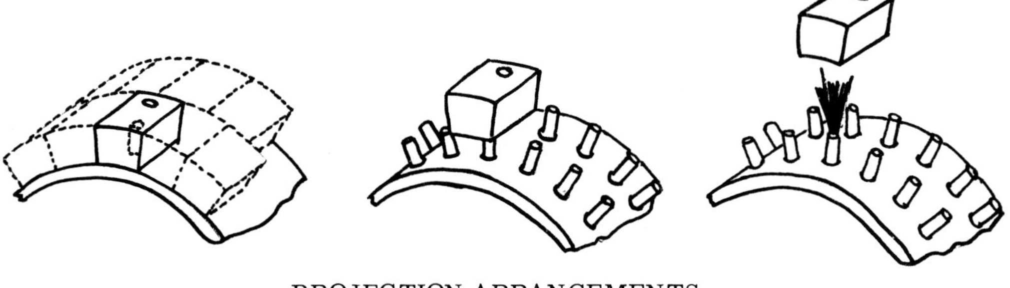

PROJECTION ARRANGEMENTS

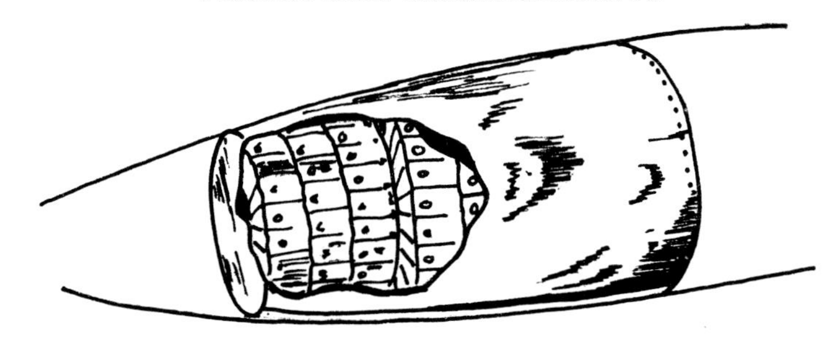

SUB MISSILE ARRANGEMENT

Fig. 7. Sub missile warhead

EXPLOSIVES USED IN WARHEADS

Properties of explosives dictate their efficiency and use. The important properties are Sensitivity, Power and Velocity of Detonation. Details of these will be found in Chapter 4.

SAFETY AND ARMING MECHANISMS

General Requirements

Although it is generally accepted that the payload of a missile comprises a warhead and fuze, it is not well known that there also exists an additional and important device known as a safety and arming unit or mechanism. This really performs the function that a traditional fuze in an artillery projectile performs. Missile fuzes are generally electronic devices whilst safety and arming systems carry the initiatory explosive components and safety devices.

The specific role of the safety and arming system is to ensure the safety of the warhead throughout its service life up to the time after launch when the missile needs to be armed. Having achieved this it has to ensure that the explosive train is then lined up ready to function on receipt of the appropriate signal. Finally it must respond to the signal and ensure timely and efficient detonation or operation of the warhead.

Forces Available

The forces available to assist in the operation of safety and arming mechanisms include set back at launch, creep forward when flight is steady, little or no spin, and each is of a much lower order than traditional gun firing forces. In addition use is often made of stored energy in the shape of springs, batteries, gas pressure, and generated energy such as gas, pneumatic, electrical and hydraulic.

Safety Requirements

Safety and arming units should comply with the Ordnance Board design safety principles referred to in the chapter on Fuzes and they form an important safety function in the whole system. Some of the basics are fail-safe, visual indication of arm/non-arm situation, compatibility, no possibility of mal-assembly and so on. Often two systems are incorporated in parallel in the unit and this 'belt and braces' technique is intended to improve reliability and safety.

SELF TEST QUESTIONS

QUESTION 1 What are the desirable characteristics of a warhead?

Answer ....

....

QUESTION 2 Explain the difference between an internal and an external blast warhead.

Answer ....

....

QUESTION 3 Why is blast less effective at high altitudes?

Answer ....

....

QUESTION 4 Explain the difference between natural and preformed fragments.

Answer ....

....

QUESTION 5 Why is correct welding important in a continuous rod warhead?

Answer ....

....

QUESTION 6 How important is beam width when dealing with missile and target matching?

Answer ....

....

QUESTION 7 Explain the difference between a bomblet and a sub projectile in a warhead.

Answer ....

....

QUESTION 8 What are the main properties required of a high explosive filling for a warhead?

Answer ....

QUESTION 9 Define a safety and arming mechanism.

Answer ....

....

....

QUESTION 10 Comment on the forces available for the design of safety and arming mechanisms.

Answer ....

....

....

ANSWERS ON PAGE 252

8

The Attack of Armour

INTRODUCTION

Armoured fighting vehicles, and in particular the main battle tank, form one of the most important, and at the same time most difficult, groups of targets to be defeated on the battlefield. Effective anti-armour ammunition is an absolute necessity for an army since its ability to defeat armoured vehicles is of fundamental importance to its success in battle.

THE TARGET

The tank has three main characteristics: mobility, firepower and protection. From the attacker's point of view protection presents the principle problem to be defeated, because this protection not only has to be penetrated but sufficient energy must remain to damage the crew, the machinery, the equipment and the other vital components enclosed within this protection. It is important to reiterate this essential point right at the start; penetrating a tank's protection is not enough to defeat a tank in most cases, sufficient energy must be available after penetration has been achieved to disable the vital components within the tank.

The tank, therefore, poses a formidable target to the projectile designer. To defeat it, a concentrated very high energy form of attack is required. This can be derived from either a kinetic energy source, or a chemical energy source, or from a combination of the two sources. As a target, the tank, if it were just a square, homogeneous, box-like structure, would be comparatively simple to defeat. It is, however, an extremely complex target to attack, as it is protected by armour plate of varying thicknesses, plate material and layout, arranged at varying angles. There is a wide variation in its armoured protection surfaces which also often include external discontinuities like spare track links, tools, lifting plugs, stowage bins and numerous other accoutrements. Thus rarely does a smooth flat plate, albeit at an angle, present itself to the attacker. The target is not a plate-type structure in any case, as it includes numerous other variations as well: tracks, wheels, track rollers, sprockets, suspension units and other components.

The angle of attack between the axis of the projectile and the surface under attack is a tri-planar consideration (between azimuth and elevation angles of attack and the angle at which the plate under attack is sloped) and will vary considerably from engagement to engagement. The target rarely exposes itself completely, is often in motion, and it usually retaliates promptly!

DAMAGE LEVELS AND ASSESSMENT

The ideal result to be achieved in the attack of armoured vehicles should be the complete destruction of the vehicle and its crew so that neither can be used again. As explained in Chapter 2 however, destruction, apart from being difficult to achieve, particularly against a complex target like a tank, is not a cost-effective criterion against which to design projectiles. The level of damage to be achieved against a tank, therefore, is one of disablement by attacking its firepower, its mobility or its crew, rather than attempting to destroy it completely. First of all though, some sort of standard or yardstick is needed so that the performance of munitions to achieve the required level of damage against armoured vehicles can be measured and compared. It is easy to express subjective opinions about the damage which a particular projectile achieves against a particular target. Subjective views, however, are not valid in any scientific analysis. Thus tank damage assessment criteria have been introduced for use as a basis for making a quantitative assessment of the effectiveness of anti-armour projectiles. The first criterion is the M or Mobility Kill which specifies that the tank is immobilised and incapable of executing controlled movement, and is irreparable by the crew on the battlefield. The F or Firepower Kill occurs when the main armament is put out of action, either because the crew has been rendered incapable of operating it, or because the associated equipment has been damaged making the gun inoperative and irreparable by the crew on the battlefield. The K Kill is the ultimate, the tank is destroyed, knocked out, immobilised and damaged beyond repair. For the attack of armoured personnel carriers (APCs) or mechanised infantry combat vehicles (MICVs) there is an additional criterion, the P Kill. The measure of effectiveness here is the percentage of the payload (the soldiers it is carrying) that are incapacitated and thus incapable of fighting on effectively.

Although it would be convenient to fire at live tanks to assess the effects of particular anti-armour projectiles, such a trial would be extremely costly and would require large numbers of highly trained staff to assess the results. Usually, therefore, simulated tank targets are used during much of the testing and development trials of ammunition. Firings against actual tank targets, with articulated dummies representing the crews, are done during the final assessment of a design, to establish whether or not the munition under test meets the specified operational requirements. The use of simulated tank targets during the initial stages of development does allow the munition to be tested under carefully controlled and reproducible conditions, so that design variations can be meaningfully compared. Simulated tank targets are nothing more than certain configurations of armoured plate arranged to represent tank targets likely to be met on the battlefield. Figure 1 shows one such configuration: the three target plates representing the possible shot path through the side of a tank, with the first plate simulating a tank's skirting plate, then a gap before the next plate which represents a road wheel or suspension unit, another gap and finally the hull plate.

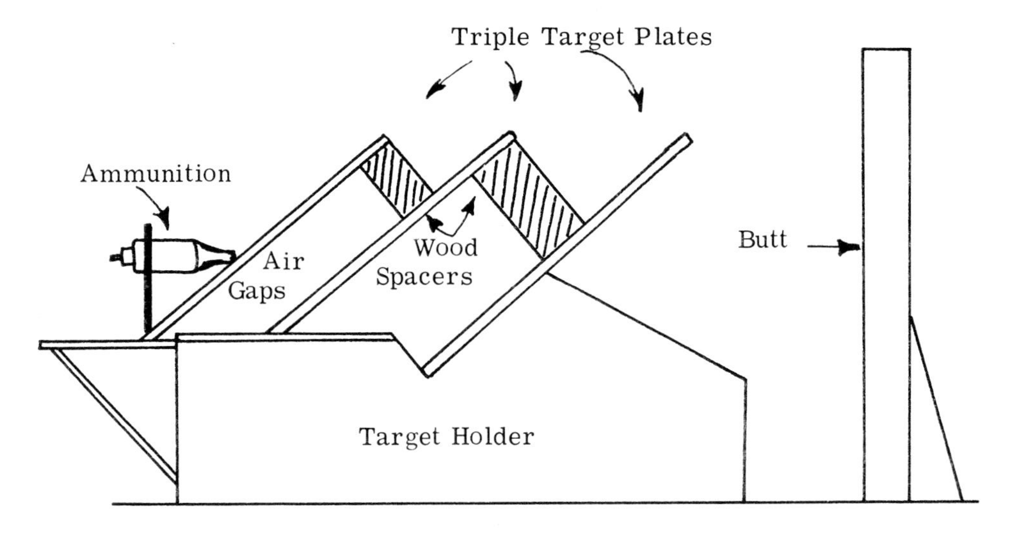

Fig. 1. Triple plate target array

Angle of Attack

In the United Kingdom, "the angle of attack" referred to in the context of attacking armour is the angle between the path (line of arrival) of the projectile and the "normal" angle to the plate under attack; angle θ in Fig. 2. The "normal" angle is itself defined as being at right angles (90°) to the target plate. The "angle of attack" of armour is the cause of some confusion in discussions between different countries on this subject, as some armies understand this to be the angle between the plate and line of shot arrival; angle α in Fig. 2.

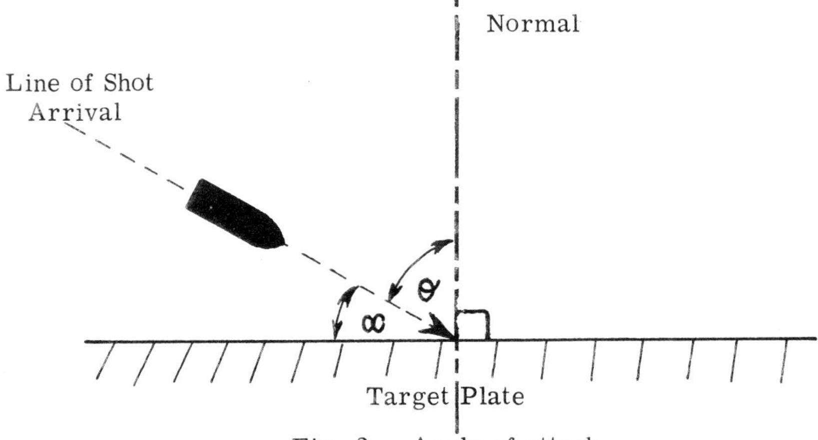

Fig. 2. Angle of attack

KINETIC ENERGY

Introduction

The kinetic energy form of attack is a sheer brute force approach whereby a solid projectile ("shot") is hurled at a tank as hard and as fast as it is possible to do so. The word "shot" is used to differentiate this type of solid projectile from a "shell" projectile which carries something to the target. A kinetic energy shot is a solid mass of dense metal, carefully shaped to give a good penetrative performance at the target. It does not have a fuze, but usually contains a built-in tracer at its rear, so that its progress can be observed by the firer. It is often referred to as an armour piercing shot (AP Shot) or solid shot. The principle of this form of attack is to hit the target with a shot which possesses a high amount of kinetic energy concentrated over a small area. If m is the shot mass, v its striking velocity and d its diameter, then the requirement is for the shot to have a high value of mv^2/d^2; in other words the shot should be a long thin pencil-shaped projectile. The complication with the kinetic energy form of attack, however, is that there is a conflict between this requirement for shot shape, size and mass at the target, and the shot requirements in flight and in the gun. Furthermore, as all the energy the shot possesses at the target has to be developed and imparted to it in the gun, there are a number of penalties to be paid for this. A large and heavy gun is required, since the size and weight of the gun are proportional to the energy imparted to the shot. A large, heavy gun may have to be mounted in a heavy structure to absorb the considerable recoil energy, and the rate of barrel wear is high compared with other guns firing lower velocity projectiles. On the other hand there are considerable advantages in having a high velocity gun, particularly for direct fire engagements. The higher velocity given to the shot means a fast, flat trajectory giving an increased chance of a hit and a quicker engagement and response time.

Target Effects

When the shot punches through the armour plate it tends to break up and fragment as well as producing fragments from the plate under attack. These fragments, particularly those from the shot itself, will be very hot, and will ignite propellant charges, fuel and possibly cause the premature function of any chemical energy projectiles that are hit within the tank. Such is the power of the kinetic energy attack that given a favourable strike the target can be disabled by causing structural damage to it. For example, the kinetic energy possessed by a 120 mm Armour Piercing Discarding Sabot (APDS) shot compares with that possessed by a 10 ton truck travelling at 70 mph, or a 50 ton tank travelling at 30 mph.

Penetration

The penetration of armour plate is a complex phenomenon. A detailed explanation of changes in plate structure and behaviour when subjected to the high rates of strain, stress and pressure imposed on it by kinetic energy attack, however, is not required for the purposes of this chapter. It is sufficient to appreciate that the mechanism of penetration is an involved metallurgical process.

It has already been established that the shot should have a high amount of energy concentrated over a small area (a high mv^2/d^2 value) to achieve penetration. To relate this shot requirement to the thickness of armour plate (T) it can penetrate at the normal angle (a flat vertical plate, struck at 90°), the formula:

$$\frac{T}{d} = \frac{mv^2}{d3}$$

is used to enable the penetrative performance of the shot to be calculated. This formula is referred to as the "fundamental armour equation", as it shows the thickness of armour penetrated per unit calibre of shot used.

The penetrative performance of shot at angles of attack other than at the normal, however, is a more complicated calculation. At first glance, it would appear that the shot should follow the cosine law as shown in Fig. 3.

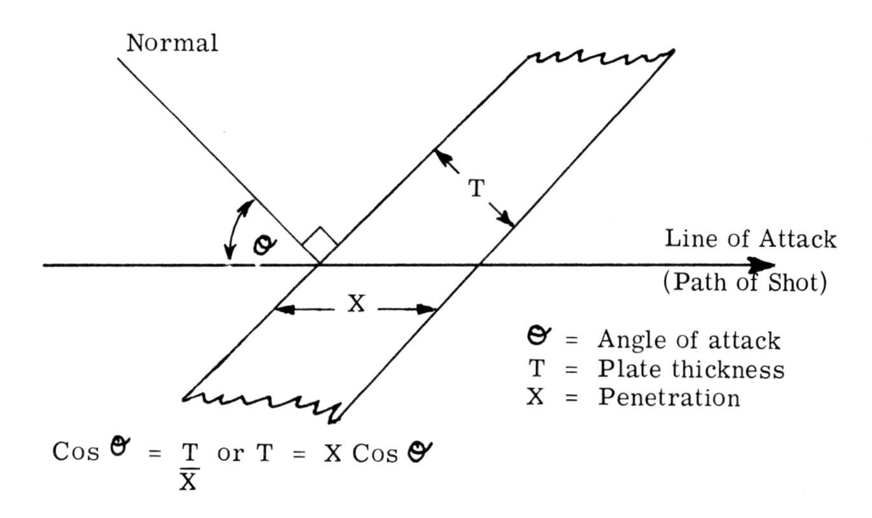

Fig. 3. Cosine law as applied to sloped armour

In fact, the shot does not follow the cosine law when it strikes at an oblique angle. It forces an S-shaped path through the armour rather than passing straight through it. A number of complex formulae, of which the best known is the Milne-de-Marre formula, can be used to calculate the penetrative performance of an angled shot strike. These formulae indicate that the penetrative performance will be worse than the cosine law, whilst in practice, modern designs of shot produce penetrative performances that are slightly better. For example, a swivel nose cap will give the shot a better performance than the cosine law. As shown in Fig. 4, a shot when striking at an angle tends to turn away from the plate and start to ricochet. When fitted with a swivel nose cap, the nose cap performs this initial ricochet movement on striking, and in doing so it forces the shot to turn towards the normal angle before starting to penetrate.

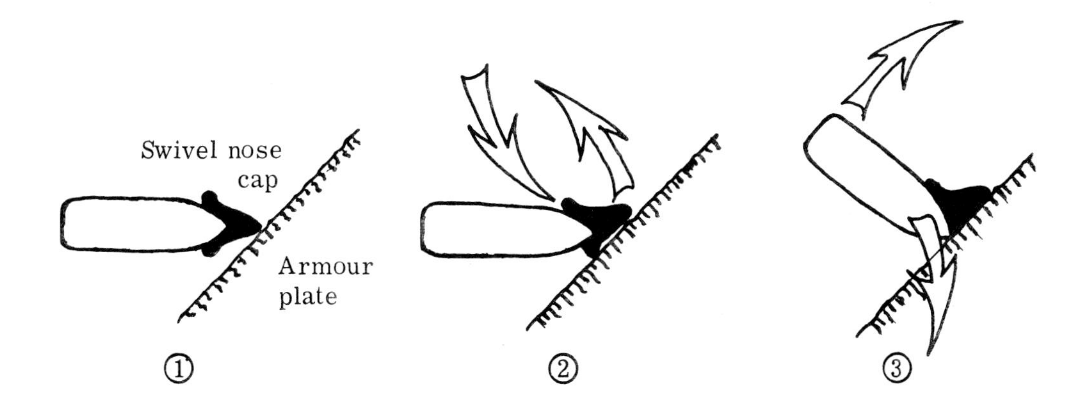

Fig. 4. The swivel nose principle

Penetrative Path of the Shot

When a shot strikes an armoured plate at a normal angle of attack it passes straight through the plate. As the nose of the shot starts to penetrate, the metal deforms and forms a "collar" or "front petalling" as in Fig. 5.1 below. Once the shot is well into the plate, the plate is subjected to further deformation as shown in Fig. 5.2, due to the plate being compressed at right angles to the shot path, and also deforming axially causing the back of the plate to bulge. Finally, the plate fails either by "plugging" (Fig. 5.3) or by plastic flow, or by "discing" (Fig. 5.4), or by a combination of both.

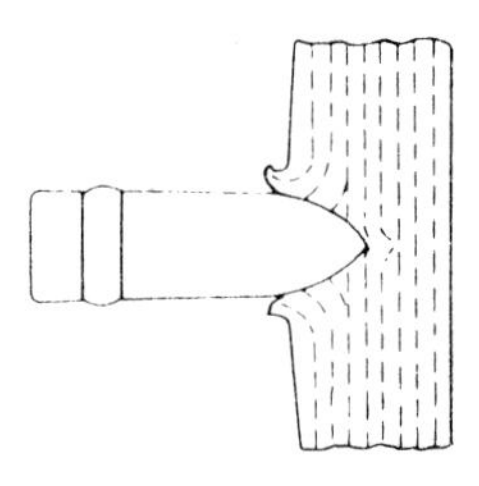

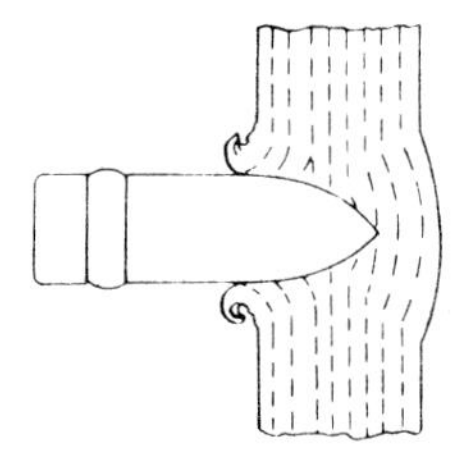

5.1 Initial penetration

5.2 Back bulge forming

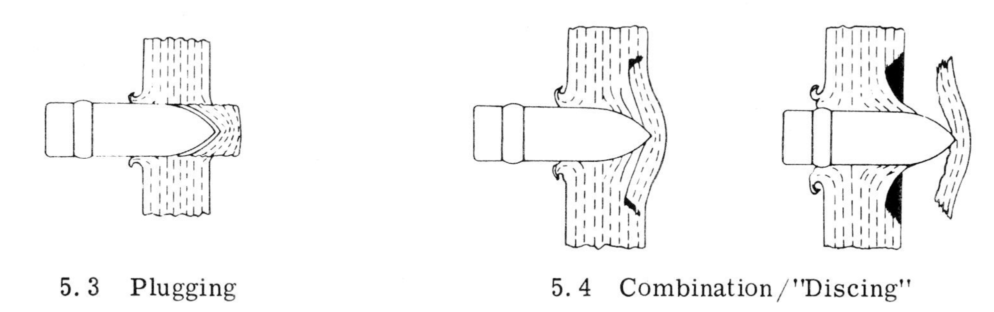

Fig. 5. Penetrative path of shot at normal angle

At an angled attack, that is over 20° to 30° from the normal, the shot follows an S-shaped path through the plate; it "wriggles" its way through. On striking at these oblique angles, provided that the angle is not so great as to cause a ricochet (ricochets begin to occur at angles from between 60° to 70° to the normal), the shot will start to penetrate. When the tip of the shot is well into the armour a plug begins to form and the shot takes this line of least resistance, which turns it back towards the normal. The sides of the hole it is making, however, then stop the shot from turning any more toward the normal. The important point to note is the extra distance through the armour that the shot has to travel to penetrate, compared to its path when striking at the normal.

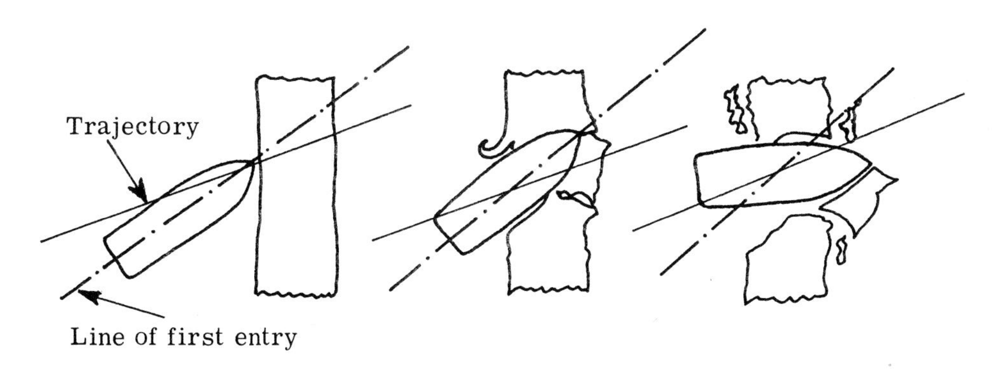

Fig. 6. Attack of armour at high angles of attack

CONFLICTING REQUIREMENTS OF SHOT DIMENSIONS

At the Target

It has been explained that a long thin dense shot is required at the target. Long thin projectiles striking armoured plate at high speeds, however, are particularly susceptible to ricochet and "break up" due to material failure. Figure 7 shows the most common ways for shot to fail when striking armoured plate.

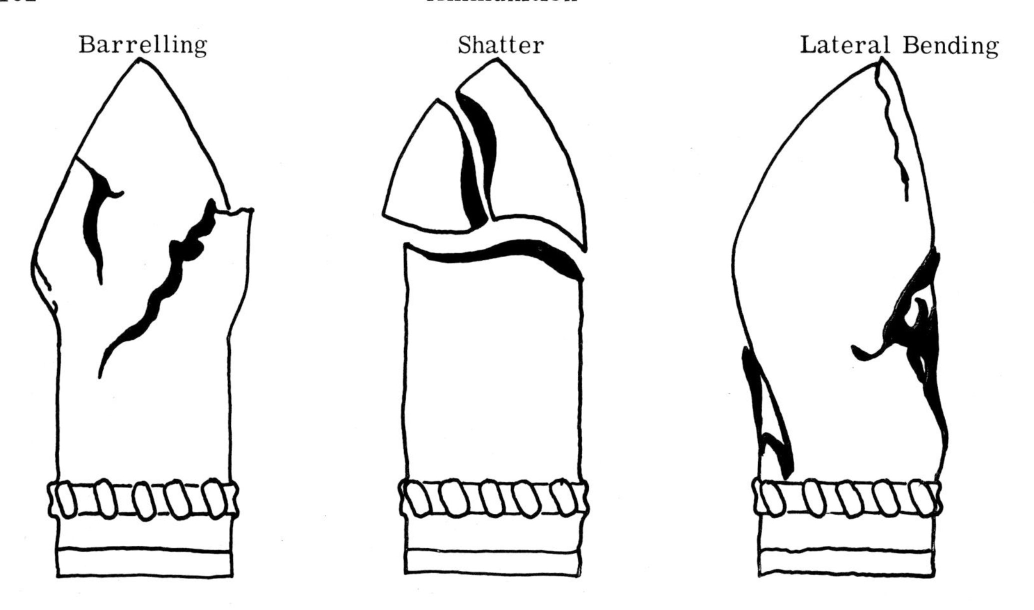

Fig. 7. Shot failures

Barrelling occurs when the nose of the shot is made of insufficiently tough or hardened metal. When it strikes, it compresses and bulges outwards increasing its cross sectional area and consequently fails to penetrate.

Lateral bending occurs when the shot hits at a high angle of attack. In this instance, it is subjected to severe lateral stresses, both shear and bending, particularly towards its rear end. The effects of these stresses can be reduced by fitting a toughened steel sheath over its rear end.

Shatter occurs when the shot strikes at the normal angle (90°) at high velocity. The shock of impact sets up stresses in the shot which it is unable to withstand, and the nose either breaks off or mushrooms. The simple shatter graph in Fig. 8, for a particular type of shot striking a plate of constant thickness at a particular angle, shows how an increase in impact velocity (or a decrease in range) can produce success or failure with and without shatter.

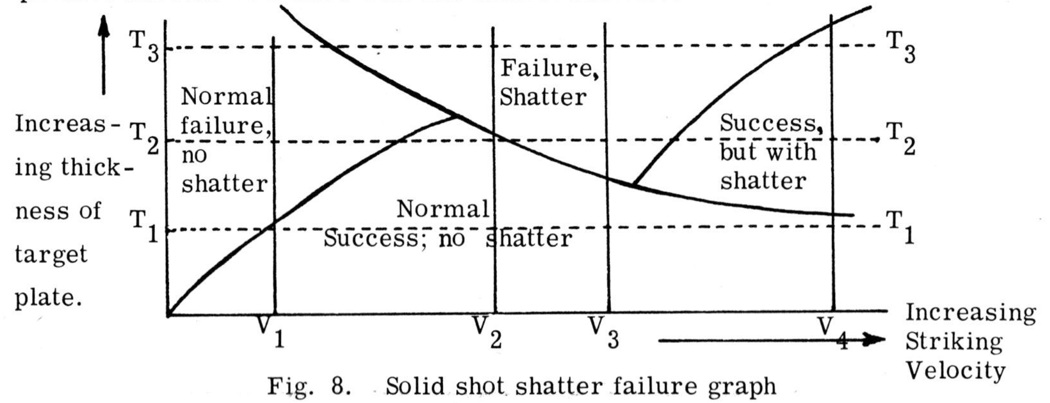

Fig. 8. Solid shot shatter failure graph

To overcome shatter a penetrative cap (of which the swivel nose cap, already mentioned, is a refinement) is fitted over the nose of the shot. This cushions the shot from the full force of impact before disintegrating and allowing the shot to start penetration. The penetrative cap has a poor ballistic shape, which affects velocity and hence range, so a ballistic cap is added. A shot fitted with a penetrative cap and a ballistic cap, is described as an Armour-Piercing Cap, Ballistic Cap (APCBC) shot. Figure 9 shows a World War II British 6-pounder APCBC.

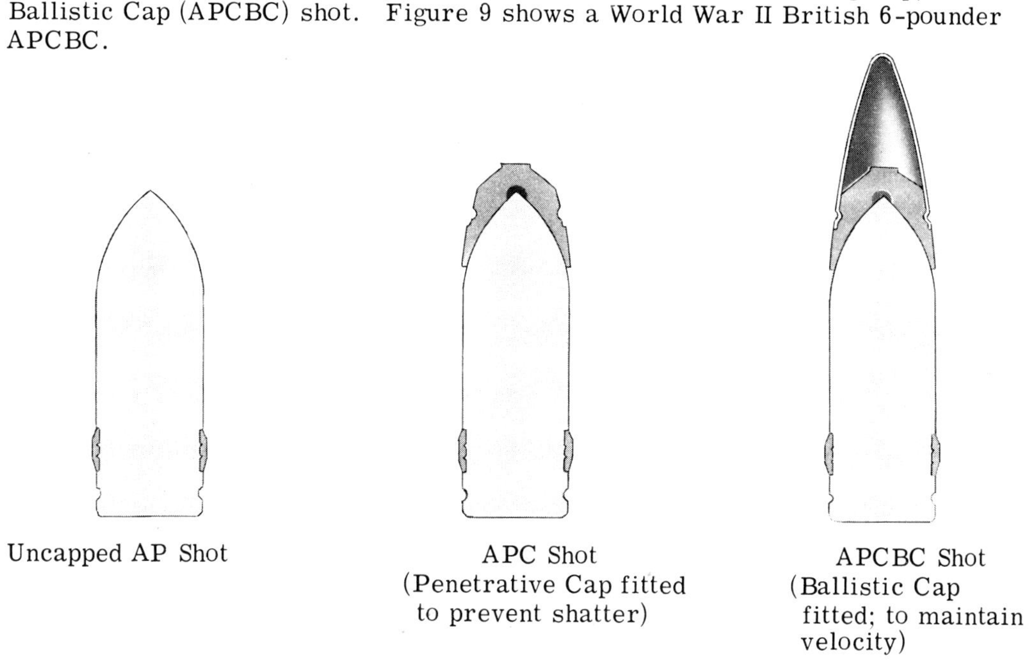

Fig. 9. World War II British 6-pounder APCBC

In Flight

To ensure the maximum kinetic energy is delivered at the target, it is important that the shot loses as little velocity as possible on its way to the target. It can be readily appreciated from the basic kinetic energy formula; $KE = \frac{1}{2} mv^2$, that the performance of kinetic energy shot is range dependent since velocity drops off with range. A dense heavy projectile has a better "carrying" power than a lighter one, and hence will have the potential for obtaining greater range. Also, a thin projectile with a small cross sectional area will maintain its velocity better than one with a large cross sectional area. So, a long thin dense shot is the requirement for the shot in flight. Also, the longer and thinner the projectile (provided of course, that it is kept as heavy and dense as possible at the same time) the better its penetrative performance. The major difficulty with long thin projectiles is that they lack ballistic stability in flight, and have to be stabilised either by spin, or by adding fins or by flaring the rear end of the projectile in some way.

Provided a shot's length to diameter ratio is less than 5 to 1, it can be spun sufficiently in the barrel of a rifled gun to be stable in flight. Where this length to diameter ratio is greater than 7 to 1, however, it is not possible, using a rifled barrel of practical length, to impart sufficient spin to stabilise the shot. To stabilise a shot with a high length to diameter ratio, drag has to be applied to its rear end. This is necessary to keep the rear end of the shot from trying to overtake the nose, due to the action of air resistance retarding the nose in relation to its tail. This requirement for aerodynamic drag can be met by fitting fins to the rear end of the shot.

As an aside, it is interesting to note the "return" of modern anti-armour projectile designers to the "anti-armour projectiles" used previously in history. For example, the English long bow arrow used so successfully by Henry V's archers against French knights at Agincourt on the 25th October 1415, was essentially a finned long rod penetrator.

Fin stabilised shot, however, presents problems if it has to be fired from a rifled gun. The designer has to include features like slipping driving bands to prevent spin being imparted to the shot as it moves up a rifled barrel. In fact, a limited amount of spin (40-100 rpm) is imparted to fin stabilised projectiles as it helps maintain accuracy. Fin stabilised shot loses velocity quicker than spin stabilised shot and is more susceptible to cross wind effects.

Requirement in the Gun

As the gun imparts the high velocity to the shot, ideally the shot should present the largest possible cross sectional area against which the propellant gases can act, and it should be light. The requirement, therefore, is that the shot should have a high value of d^2/m: it should be a short squat projectile made of a low density material. This is exactly opposite to the requirements in flight and at the target. Although at first glance, a lighter shot appears to be contrary to what is needed elsewhere, it should be remembered that kinetic energy is equal to the product of half the mass times the velocity squared. So if some mass is sacrificed to obtain an increase in velocity, there is still an overall gain in kinetic energy.

Reconciliation of Conflicting Requirements

The differing requirements for shot shape, size, and weight in the gun, in flight and at the target have been reconciled in the type of shot described as Armour Piercing Discarding Sabot (APDS) shot. The shot can be made relatively lighter by enclosing a small high density sub calibre core, within a larger diameter full calibre envelope, called a sabot. This sabot is discarded shortly after leaving the muzzle of the gun, leaving the small sub calibre heavy core to continue to the target. The core material is made of a high density material such as tungsten; tungsten having the same order of specific gravity as gold, but rather less than platinum! Other high density materials such as depleted uranium are also used. Figure 10 shows the make up of a modern APDS shot. Also shown is a Soviet 115 mm Armour Piercing Fin Stabilised Discarding Sabot (APFSDS) shot. There is a trend towards fin stabilised shot, because of the increased penetrative performance to be gained from shot with high length to diameter ratios.

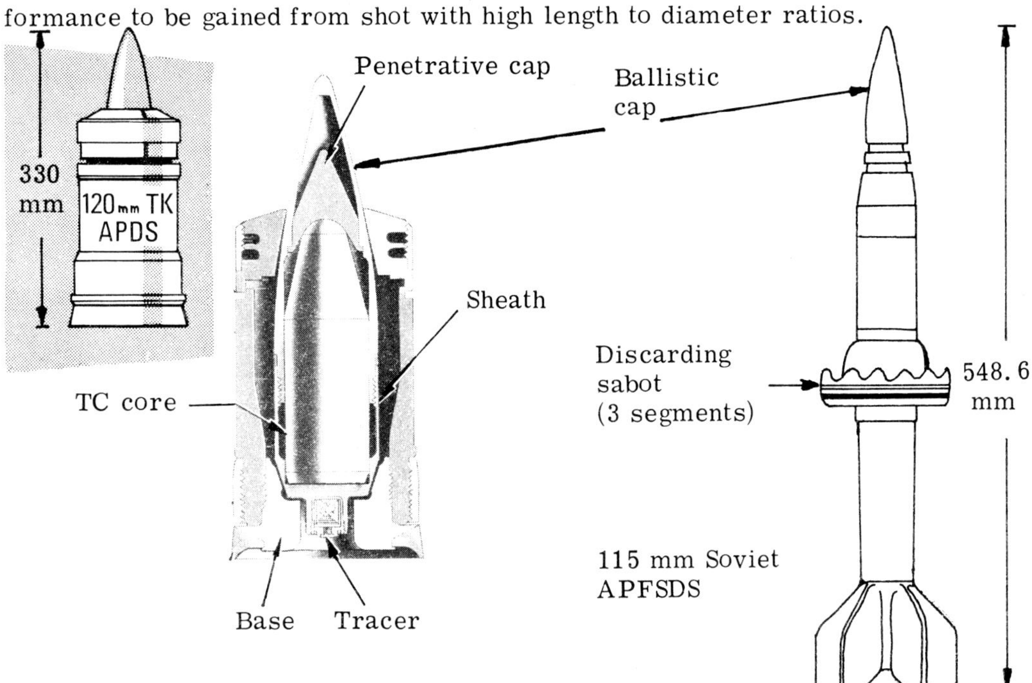

Fig. 10. Discarding sabot shot

The sabot which encases the sub calibre core in the gun barrel is so designed that it breaks up as it moves along the gun barrel. This does mean of course that discarding sabot shot cannot be fired through barrels fitted with muzzle brakes, for obvious reasons. Within the barrel, the sabot constrained by the barrel remains around the core. On leaving the muzzle, the sabot can be discarded in one of two ways depending on its initial design. A "petal" sabot, usually three segments rather like, as its name implies, three flower petals, which are discarded radially due to the spin of the shot. Alternatively, a "pot" sabot can be used which relies on air resistance retarding the sabot axially away from the core. Figure 11 shows the sabot discarding from an APDS round.

11.1 Discarding petals

11.2 Petal sabots clear

Fig. 11. Armour Piercing Discarding Sabot (APDS) Shot.

The Evolution of Modern AP Shot

The evolution of modern anti-armour kinetic energy shot is fascinating and is worth looking at briefly. It clearly shows how designers have wrestled with the problems of trying to reconcile the conflicting requirements for shot dimensions, before arriving at the discarding sabot solution.

The Germans began the process in World War I by firing a specially hardened small arms 7.92 mm bullet propelled by the charge used in a 13 mm machine gun round, against tanks. This large charge weight gave the small bullet an increased velocity with which to penetrate armour plate.

An early purpose-designed AP round was the British "2-pounder" steel shot. This retained the principle of a high charge weight to achieve a high velocity, with additional design refinements to improve performance. The shape of the head of the shot, and the application of a hardness gradient along its length and across its width helped to minimise ricochet and defer shot material failure.

Evolution continued with the addition of penetrative caps and ballistic caps, together with larger calibre designs consequent to improvements in gun design.

The "6-pounder" Armour Piercing Composite Rigid (APCR) shot was the next significant development. The thicker armours on World War II tanks meant that anti-armour shot required even higher velocities, certainly above 1000 metres per second, to defeat them. Full calibre solid shot was heavy and to achieve the necessary velocities, an increased charge weight and a longer gun barrel were required. This resulted in high levels of barrel wear and unwieldy guns. The APCR shot was an attempt to "lighten" the shot, by giving it a light alloy outer envelope containing a small tungsten carbide core. The full calibre, relatively lighter, shot could attain the high velocities required, but because it was full calibre it lost velocity very quickly and was, therefore, only effective over short ranges. Thus the major drawback of the APCR shot was its poor value of m/d^2 in flight. Furthermore, since only the tungsten carbide core penetrated the target, much of the energy imparted to the full calibre shot, which merely "carried" the core to the target, was wasted.

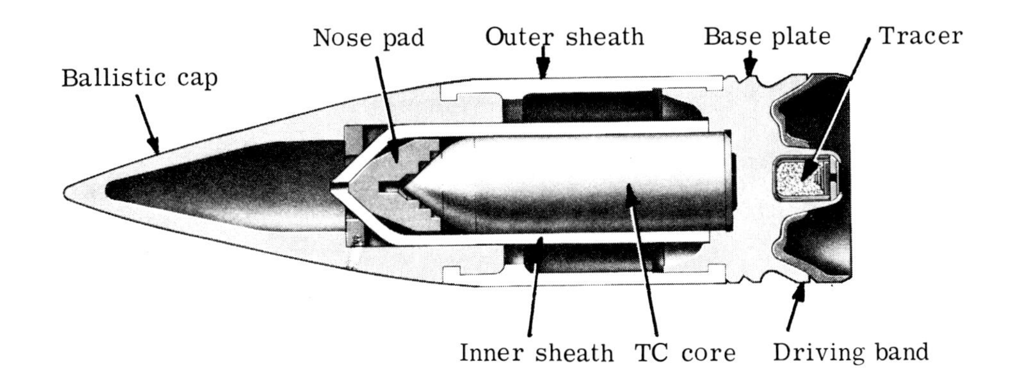

Fig. 12. Armour Piercing Composite Rigid (APCR) shot

An ingenious attempt was then made to try and overcome the disadvantages incurred by retaining a full calibre for the shot throughout. This solution involved reducing the calibre of the shot for flight and saw service as the Armour Piercing Composite Non Rigid (APCNR) shot shown in Fig. 13.

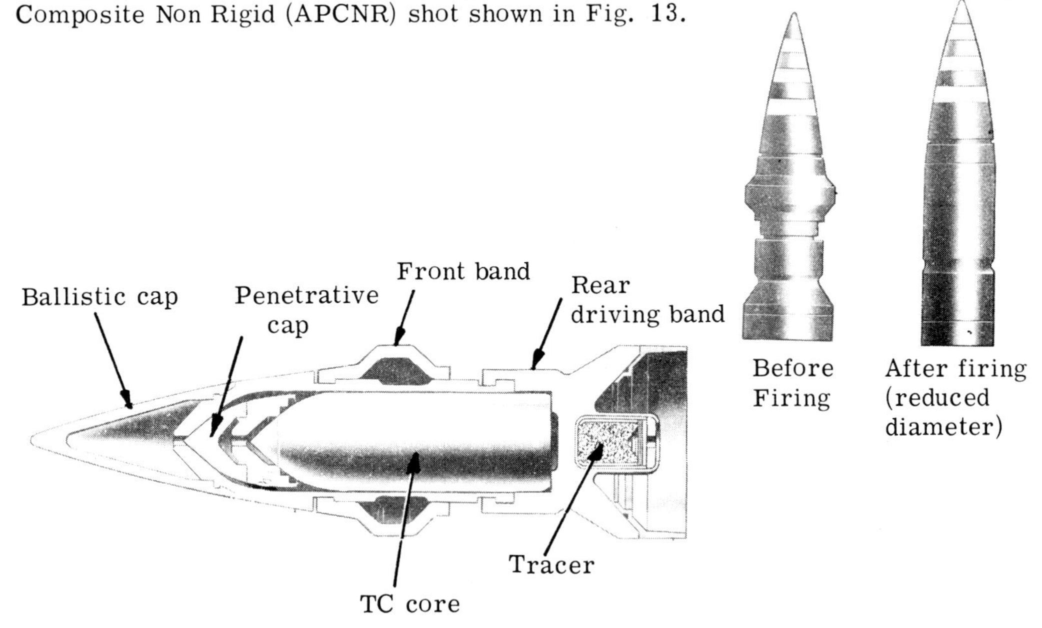

Fig. 13. Armour Piercing Composite Non Rigid (APCNR) shot

The Littlejohn squeeze bore shot used a sub calibre tungsten carbide core which was carried in the gun by two deformable flanges. These flanges were swaged to the same calibre as the core by firing the shot through a tapered attachment at the end of the barrel (see Fig. 14). It was important to remember to remove the tapered adapter if a projectile other than a APCNR shot was to be fired next!

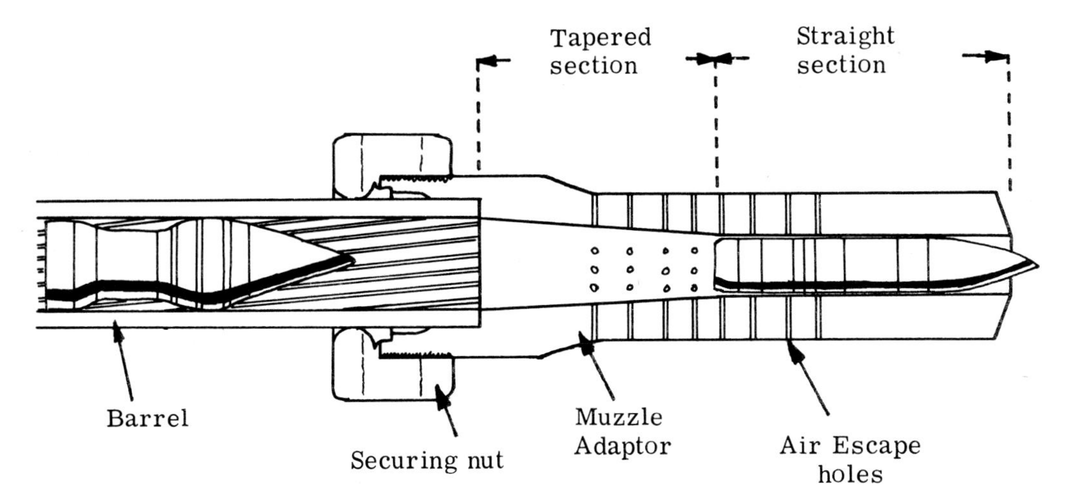

Fig. 14. Littlejohn squeeze bore adaptor

Thus was the modern discarding sabot anti-armour shot (APDS and APFSDS) evolved. They represent the latest solutions to defeat the tank; a process which began with the appearance of the first tank in 1916.

The range of alternatives available for increasing penetration is shown in diagrammatic form at Fig. 15.

Summary

To summarise, the AP kinetic energy shot is a potent attacker of armour, with a high chance of a hit and a fast engagement and retaliatory capability. It is a method of attack used by all the armies in the world with sizeable armoured forces.

CHEMICAL ENERGY

Introduction

There is a certain sophistication about the chemical energy approach to the defeat of armour. It involves delivering the chemical energy to the target and then initiating it with a fuze. Chemical energy can be utilised to attack armour in a number of ways, but not all are equally effective. The use of straight forward high explosive shell with its blast and fragments effect is not really effective against tanks, unless a direct hit with smaller calibre shell or a very near miss with large calibre (155 mm and above) shell can be achieved. If this can be achieved, HE can have a considerable effect on tanks; radio aerials, periscopes, sights and tracks can be broken and damaged. Furthermore, the effect on the tank crew of a large quantity of HE detonating on, or close to, what is in effect a metal box, can be imagined. HE shell, however, is designed primarily to attack people and only secondarily material it is not designed to attack armour.

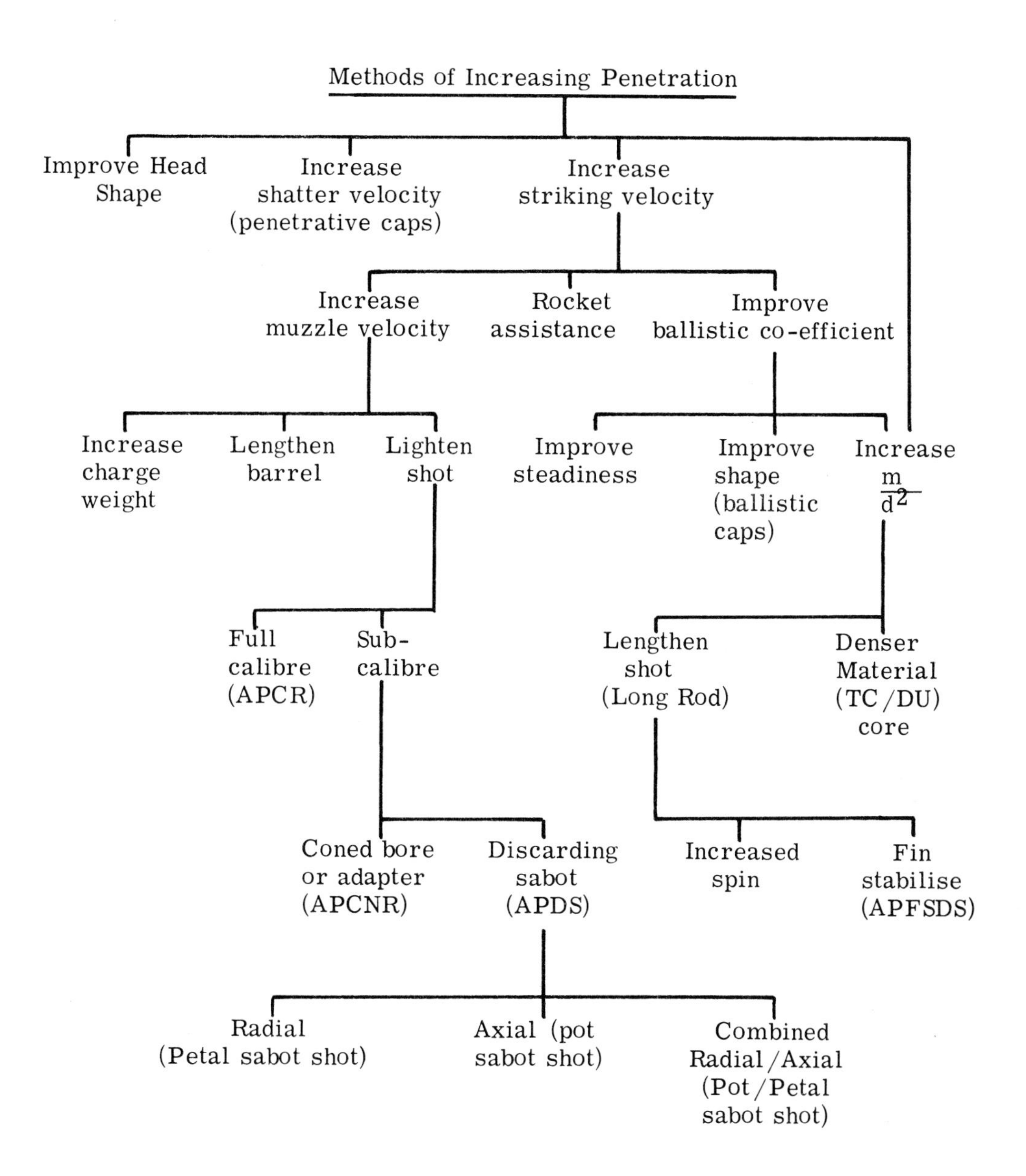

Fig. 15. Methods of increasing penetration

A - I

The defeat of main battle tanks requires purpose-designed chemical energy projectiles if an adequate level of lethality is to be achieved. The chemical energy methods of attack most used are the shaped charge, squash head and plate charge effects. These are described in detail below.

HIGH EXPLOSIVE ANTI-TANK (HEAT) EFFECT

High Explosive Anti-Tank (HEAT) (which does not rely on heat to achieve its effect, incidentally uses the hollow charge, or shaped charge, or Munroe or Neumann effect. The various names by which it is known relate to its method of operation or to the people who developed the concept. Figure 16 shows the development of the effect. Initially, it was observed that when a high explosive was detonated in close contact with an armoured plate a small indentation was made in the plate. If the explosive charge had a wedge-shape cut into it, thereby concentrating the detonating wave, an even larger indentation could be achieved. The effect could be further amplified by removing the charge a suitable distance ("stand off" distance) from the plate. Lining the wedge-shape cut in the explosive with a malleable metal increased the penetration of armoured plate even more dramatically.

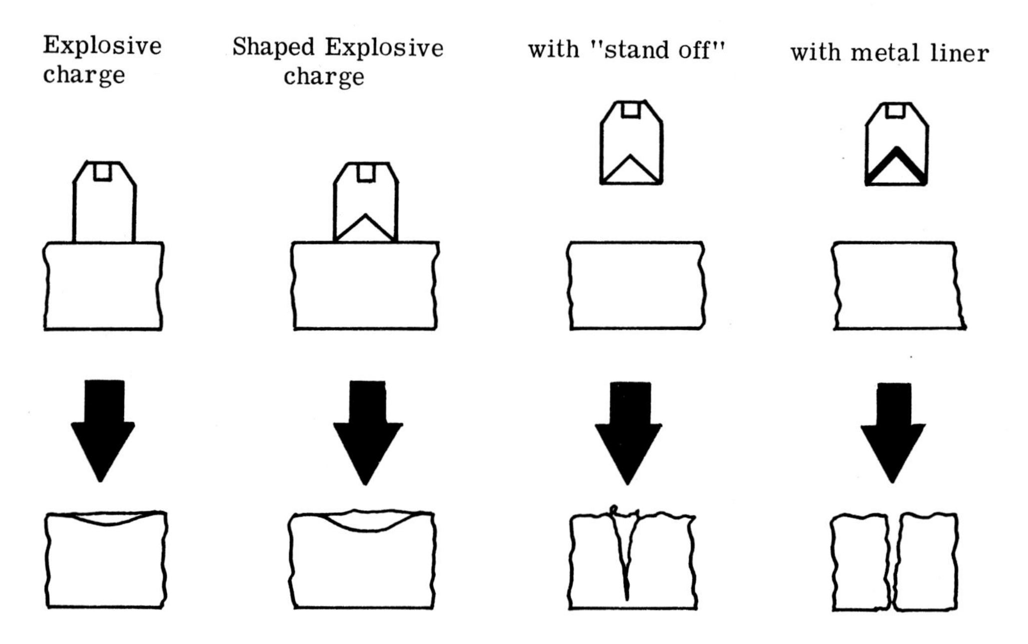

Fig. 16. The development of the HEAT charge

Concept

In principle, therefore, HEAT works by using the energy available from the detonation of a charge of high explosive to collapse and break up a metal liner into a metallic jet and plug. As projectiles are cylindrical the liner in a HEAT shell is cone shaped, and is often referred to as "the cone". About 20% of the metal liner goes into the metallic jet which has a velocity gradient from its tip of approximately 8000 to 9000 metres per second, to its tail of about 1000 metres per second. The remaining 80% of the liner forms a plug which follows the jet at a much lower velocity in the order of 300 metres per second. Figure 17 is a representation of what happens.

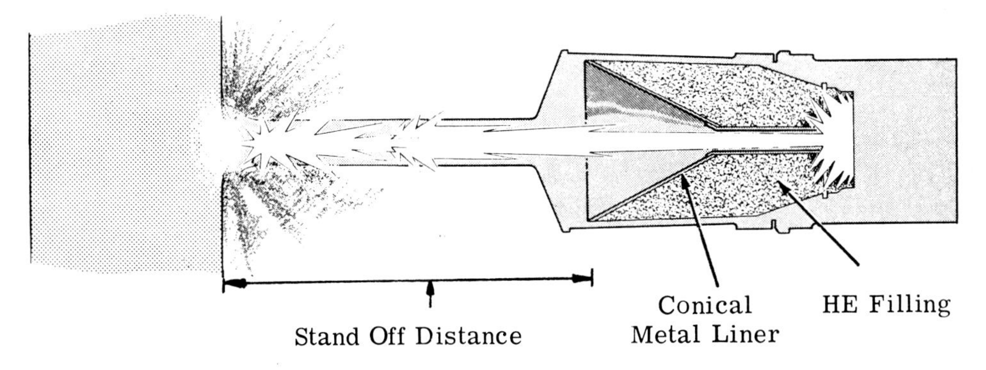

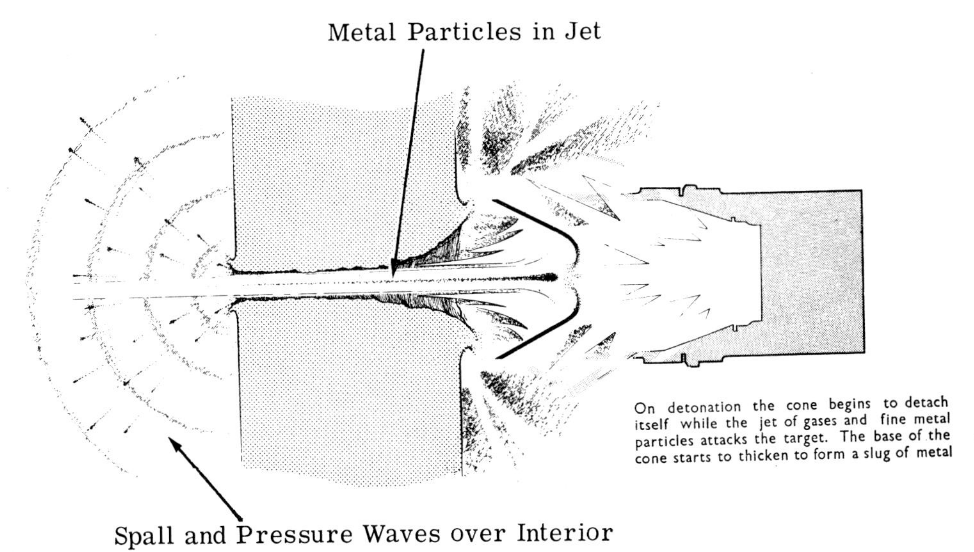

Fig. 17. HEAT effect

Penetrative Performance

The jet achieves its penetration solely by the intense concentration of kinetic energy at its tip which exerts a pressure of some 200 tons (308 mega pascals) per square inch on the armoured plate. Under this intense strain the plate gives radially; it simply tries to get out of the way. In doing so it is permanently deformed. The penetration achieved by a HEAT projectile is quite spectacular, and small (in terms of the weight of high explosive used) charges can penetrate considerable thickness of plate.

The penetration achieved is proportional to the cosine of the angle of attack; it follows a simple cosine law. The jet effectively only sees the thickness of plate directly along its line of attack (see Fig. 3), and the thickness it will penetrate, all other parameters being constant, is proportional to the size of the mouth of the cone. A well designed HEAT warhead will penetrate between three and four times the diameter of its cone.

Lethality

Whilst penetrative performance is important, it is not, particularly where HEAT is concerned, the only consideration. As in any form of attack, the essential requirement is not only to penetrate, but to have sufficient residual energy to do damage behind the plate. HEAT causes damage behind the armour in three ways; with the jet itself, with the "spall" as the debris of plate fragments and splinters caused by the passage of the jet through the plate is called, and by physiological and pyschological effects against the crew achieved by pressure, temperature and flash. The jet will, having penetrated, disable anything in its path, but as it is so narrow its chance of hitting anything inside the tank is comparatively small. The main lethality from HEAT comes from the spall introduced into the tank by the jet as it penetrates the protecting armour. The wider the exit hole on the inside of the tank, the more spall will be formed and the greater will be the lethality. HEAT lethality, therefore, is often assessed on the basis of exit hole diameter. Lethality, however, is obtained at the expense of penetration and vice versa. The narrower the jet, the greater the penetration but the less the lethality. On the other hand, the wider the jet, the greater the lethality but the less the penetration. The HEAT warhead designer always has to compromise between these conflicting factors.

Physiological and pyschological effects on the crew can arise from the over pressures which, if the jet penetrates into the crew compartment, are of sufficient magnitude to damage unprotected eardrums. Similarly, the rise in temperature can burn bare skin, and the peak level of light intensity set up as the jet passes through the compartment is sufficient to temporarily blind anybody looking directly at it. These effects can be enhanced by the choice of cone material. As the effects last for such very short periods of time, and tank crews invariably have earphones on and full overalls, any level of damage achieved by them are regarded as a bonus.

Other Factors Affecting Performance

Cone diameter is an important consideration for the performance of HEAT ammunition as lethality is affected by it, as already explained. This cone diameter/lethality relationship is not linear though, and it begins to curve and flatten out at a cone diameter of about 130 mm, as can be seen from the graph at Fig. 18.

Fig. 18 also shows that a cone diameter less than 76 mm will give poor lethality, but above this and up to about 150 mm a significant rise in lethality is obtained for every millimetre increase in cone diameter. As cone diameters are approximately 80% to 85% of gun calibre, HEAT projectiles of 155 mm plus are required to achieve good lethality.

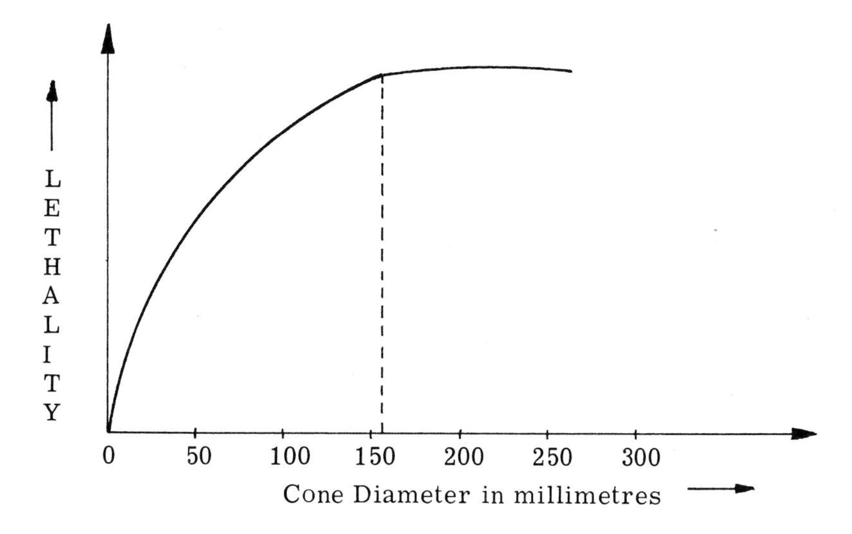

Fig. 18. Cone diameter / lethality of HEAT warheads

Spin also has a marked affect on the lethality of HEAT ammunition. As the jet is metallic, it has mass and it will be affected by centrifugal force. To achieve good penetration, however, the jet must be concentrated otherwise its performance falls off markedly. Designers have employed a range of techniques to counter the degrading effects of spin, apart from the obvious one of using a fin stabilised projectile fired from a smooth bore gun. Slipping driving bands, fluted liners, and even liners on a ball race are examples of counter spin methods employed by designers. These all increase the cost and complexity of the ammunition, and increase the potential for malfunction.

The material and thickness of the liner are other important design considerations. Copper is widely used in British land service ammunition because it deforms and flows easily to form the jet, and gives good penetration performances when compared with other metals. Steel and aluminium liners are widely used by other countries. Aluminium liners can produce an incendiary effect when penetrating, thus possibly enhancing lethality by increasing the risk of fire inside the tank after penetration has been achieved. The thickness of the liner generally varies between 1% and 5% of the cone diameter, with the best penetration being achieved by the thinnest liners.

The effect of cone shape on penetrative performance has already been touched upon. The included angle at the apex of the cone is normally between 40° and 80° and is usually 60°, although this can be varied to increase or decrease penetration (thus reducing or enhancing lethality). The cone can be truncated at its apex by up to 10% of its length without penalty to its penetrative performance.

As the jet must have distance to concentrate and stretch to achieve optimum penetration, stand off distance is a critical factor. Figure 19 shows how penetration varies with stand off. It can be seen that the peak performance is achieved at a stand off distance of about 4 cone diameters. It is almost impossible, however, to ensure this optimum stand off distance in a projectile moving at speed, so the ideal stand off is seldom achieved. For static shaped charges, such as those used by engineers for demolitions and similar tasks, precise stand off distances are easily obtained by attaching legs or stands of the precise length to the charge.

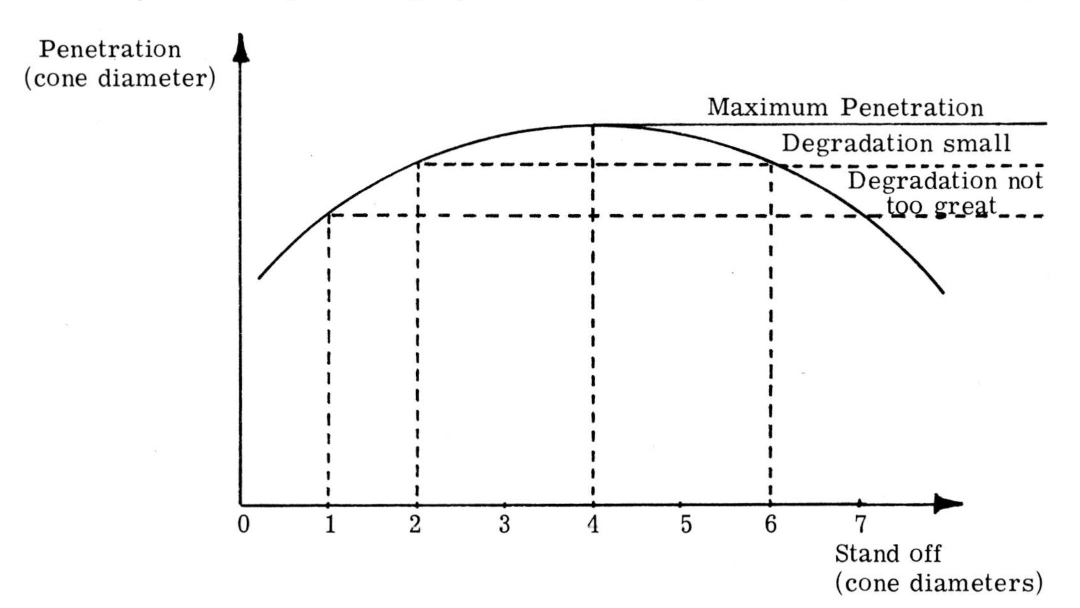

Fig. 19. HEAT attack: penetration v stand off

The usual way of trying to achieve optimum stand off distance for a projectile is to build in a nose boom or spigot containing a fuze of the "spit back" variety (see Chapter 11). A projectile with a nose boom is a tell tale sign that it is a HEAT ammunition nature, although not all HEAT natures do have a nose boom. From Fig. 20 it will be apparent that such a projectile is awkward to handle and more susceptible to damage than a normal conventionally shaped projectile.

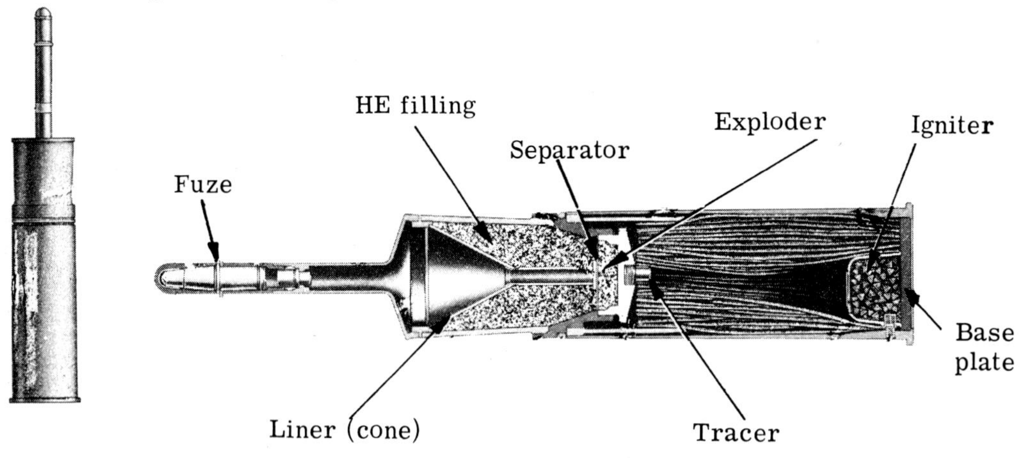

Fig. 20. A HEAT projectile

A nose fuze does marginally affect the jet's performance, as it has to pass through the disintegrating fuze before starting to penetrate. The fuze cannot be offset in some way to allow the jet free passage, because any lack of symmetry adversely affects performance by diffusing the concentration of the jet. This requirement to align the components of the warhead to the axis of the projectile to ensure symmetry, calls for close manufacturing tolerances and increases the cost of this type of ammunition.

To ensure maximum performance from the liner, the detonation shockwave should reach its full velocity of detonation as it reaches the apex of the liner. Furthermore, the wave should be as nearly planar as possible at that point. The optimum length of charge has been shown by experiment to be one cone diameter behind the apex. Increasing the charge length to three cone diameters increases penetration by only about 10%. The extra weight and cost incurred by doing this for such a small increase in performance is just not worth it for most warheads. The weight and size of the charge can be reduced by tapering the charge from the apex of the liner back towards the detonator, since the diameter of the charge at the rear does not affect the acceleration of the detonation wave. Yet again, the shape of the wave can be improved by the incorporation of multiple detonators around the periphery of the charge, or by including an explosive lens or wave shaper. (Explosive lens and wave shaper are explained in the Glossary). Again, however, these refinements add considerably to the cost of the warhead.

Finally, the apparently obvious step of increasing the weight of explosive to enhance effect does not follow in the case of the shaped charge. The weight of the charge rises as the cube of the scaling factor; in other words, in order to double penetration, an eight-fold increase in the weight of the charge is required.

Summary of HEAT Characteristics

A HEAT anti-armour projectile is capable of achieving considerable depths of penetration in armoured plate. If the warhead is small in terms of cone diameter, however, although a good penetrative performance can still be obtained, its lethality will be poor. There are recorded instances, for example, of Israeli tank crews in the 1973 Yom Kippur War not being aware that they had been hit by HEAT warheads even though their tanks had been penetrated. The lethality of HEAT is very much a function of the size of the exit hole on the inside face of the armour under attack. HEAT is not affected by spaced armour: it remains effective and goes on penetrating until its energy is used up. It is, however, sensitive to spin and the designer has to counter, somehow, the effects of spin when this type of warhead is fired from a rifled gun. Nevertheless, in terms of target effect compared to the other chemical modes of attack, HEAT is efficient and economic in its use of high explosives. For this reason it is ideally suited as a warhead for use in lightweight anti-tank weapon systems like those that have to be carried by the soldier, or for use in anti-tank guided weapon (ATGW) systems. So although for British tank guns, KE and High Explosive Squash Head (HESH) projectiles are preferred, for lightweight anti-tank weapons and especially ATGW where cone diameters of a 130 mm and more can be obtained because gun calibre does not constrain them, HEAT warheads are almost exclusively used.

HIGH EXPLOSIVE SQUASH HEAD (HESH) EFFECT

The High Explosive Squash Head (HESH) mode of attack is also referred to by the Americans as the High Explosive Plastic (HEP) method of attacking armour. The action of HESH is to detonate a quantity of high explosive in contact with armour plate, thereby sending a high velocity compressive shock wave through it (Fig. 21). When this shock wave reaches the rear of the plate, the change in medium between plate and air causes the wave to reflect back through the plate as a tension wave. When this rebounding tension wave meets further primary shock waves coming in the other direction, they combine setting up a reinforced shock wave. This exceeds the strength of the plate, and a large scab is detached from the rear surface with a considerable velocity of between 30 to 130 metres per second. A well designed HESH warhead will produce a scab about $1\frac{1}{4}$ to $1\frac{1}{2}$ times its own diameter.

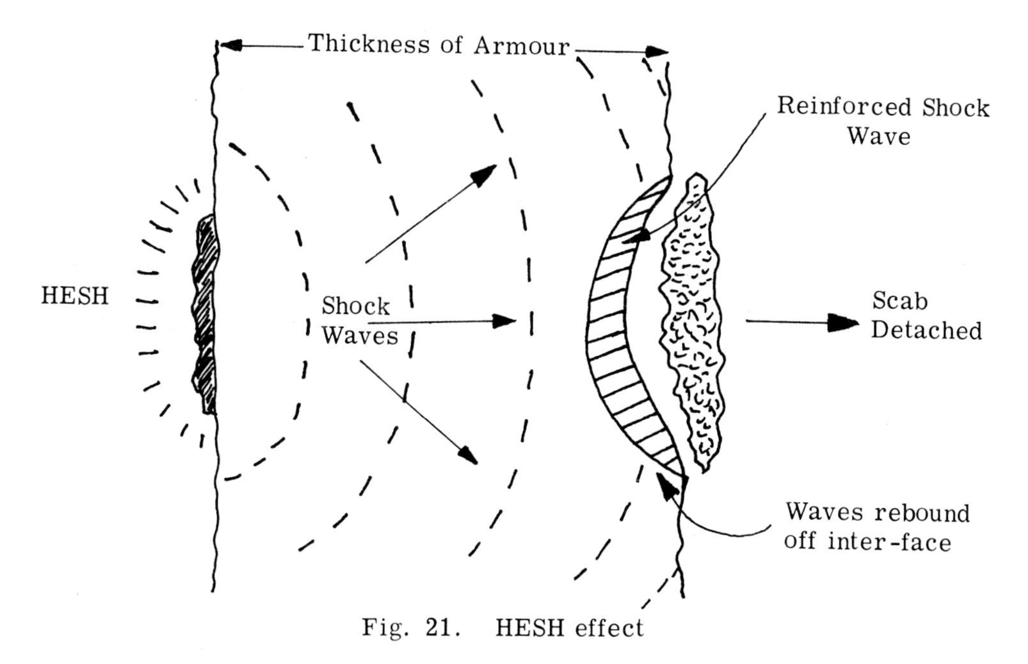

Fig. 21. HESH effect

The interesting point to note about this method of attack is that complete penetration - "perforation" - is not necessary. Until HESH was developed, it was always considered that armour had to be perforated to do damage behind it.

Main Characteristics of HESH

There are a number of factors to be considered for the design of HESH projectiles. To ensure that the optimum effect is achieved at the target, the HE filling must be in close contact with the plate under attack, and preferably spread over an appreciable area of it, at the moment of detonation. It is also necessary to ensure that the shock wave produced from the detonation is travelling towards the plate. This entails initiating the HE filling with a base fuze, to allow the HE to spread onto the target before being detonated. The shell wall must be thin to facilitate the crushing of the shell on impact and allow a good spread of the explosive on the target. The HE filling must be sufficiently insensitive to withstand impact on the plate without burning to detonation before the fuze can function to produce detonation at the optimum moment. If this preemptive detonation occurs, the shock wave travels away from the plate rather than through it. This phenomenon is known as "reverse impact detonation". Unfortunately, high explosives, even the relatively insensitive ones, are by their very nature susceptible to premature burning to detonation under these circumstances: that is being impacted onto hard armour plate at velocities in the order of 600-700 metres per second. In an attempt to overcome this problem the nose of a HESH shell usually contains a pad of inert substance (see Fig. 22) to absorb some of the impact energy and lessen the chance of reverse impact detonation occuring. Striking velocity is also limited to about 700 metres per second, which gives the HESH projectile a comparatively long time of flight to the target.

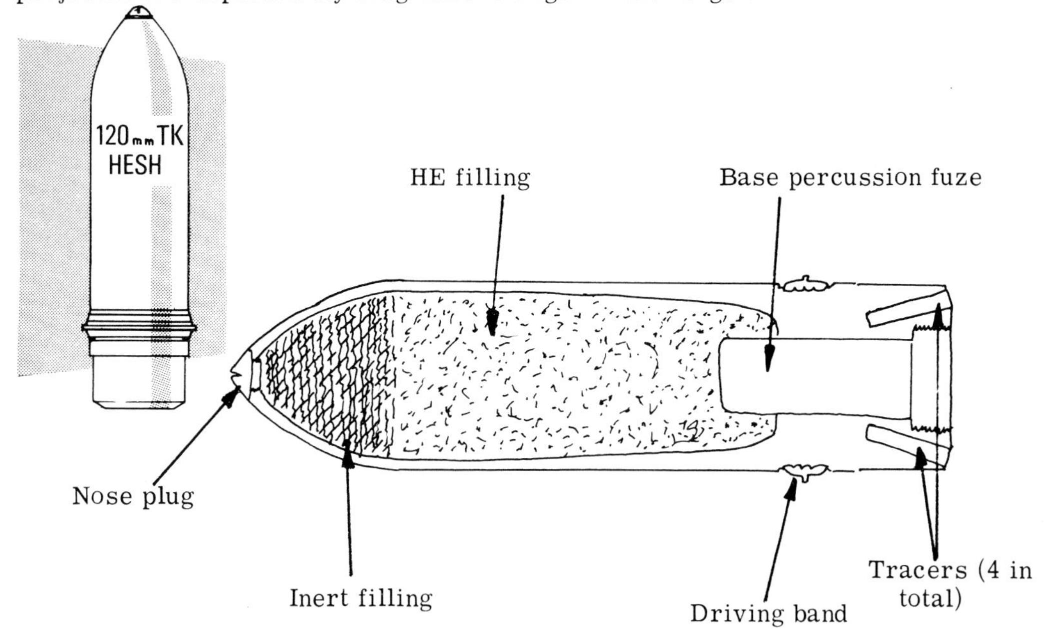

Fig. 22. A typical HESH shell

The most serious limitation of HESH is that the effect is entirely defeated by spaced armour of any sort. The shell functions on the first hard surface it strikes, and any scab it produces will be insufficient to penetrate any further armour of substance. It has been calculated that a HESH projectile striking at 700 metres per second is going to function on any plate thicker than 6 to 8 mm: anything less and its inherent kinetic energy will allow it to penetrate. It is argued that even if a HESH projectile does not produce a scab, the detonation of a weight of explosive (for example, a 120 mm tank gun HESH shell contains about 7 lbs of HE) in contact with armour plate will at least stun the crew who are, after all, sitting in a metal box. Such a detonation is also likely to produce other damage to sensitive equipment in the tank.

The performance of HESH also falls off markedly if the charge is disrupted in any way. For example, corrugated armour or any other discontinuity (lifting eyes, spare track, suspension units) which "interrupts" the spread of the HE filling, prevents a coherent shock wave, or tensile wave if the discontinuity is on the inside of the plate, forming. Unlike the kinetic energy and HEAT modes of attack, however, the performance of HESH is largely unaffected by the angle of attack. Sloped plate can help to increase the area of spread of the HE filling, see Fig. 23, although there are limits to this as ricochets start to occur at about 65° and consistent functioning falls off above angles of 60°. Figure 24 shows a more precise relationship.

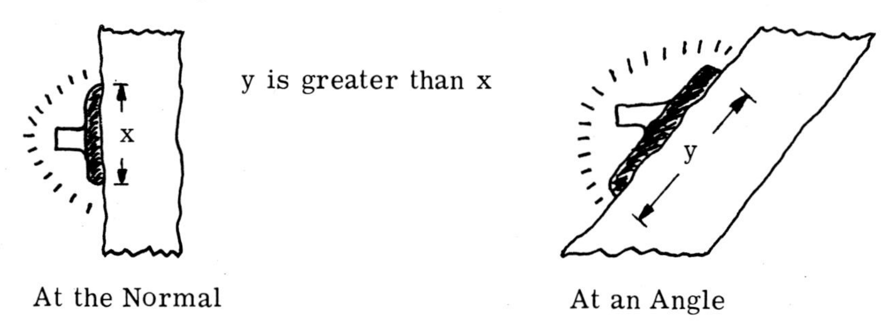

Fig. 23. Spread of HESH on vertical and sloped plate

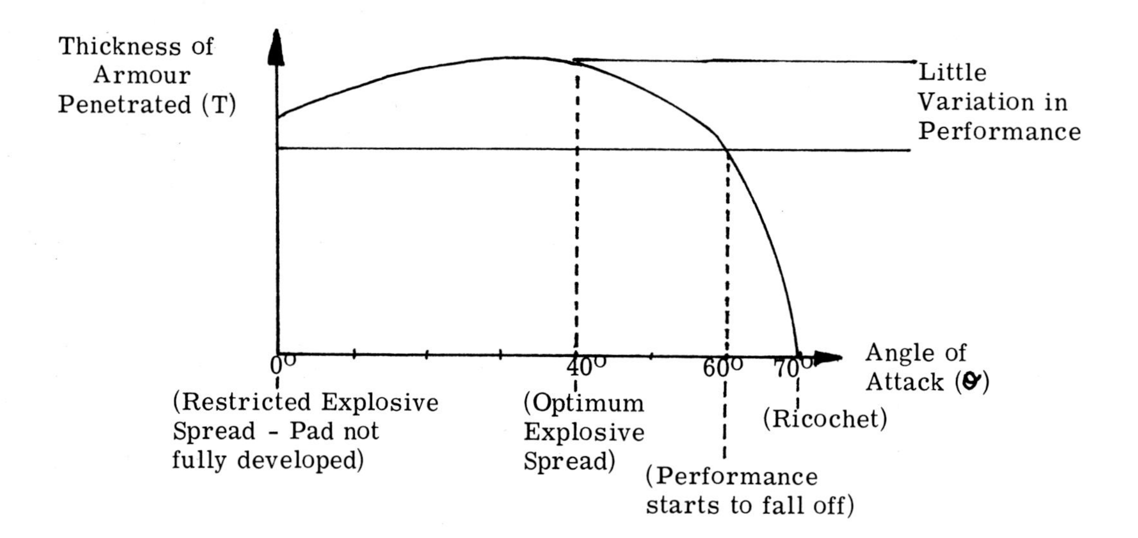

Fig. 24. HESH - penetration v angle of attack

Despite its limitations against armour, HESH is an effective multi-purpose shell. It is to all intents and purposes a high capacity HE shell and as such it has an inherent anti-personnel effect. It is also effective against concrete, and to a more limited degree against reinforced concrete, as well as against structures such as pill-boxes, emplacements, bridge supports and similar constructions. It is because of its versatility against a range of targets, that HESH is retained in British tanks as the alternative to the kinetic energy shot.

PLATE CHARGES

The other form of chemical energy attack is a cross between HEAT and HESH, and is known as the "plate charge" effect. There are various types of plate charge: the 'Miznay-Schardin plate'; the 'P (for plate) charge' itself: and the 'implosive' or 'self forging fragment' plate. Although the terms are used interchangeably, because outwardly they do not show any marked differences, they do differ very much in the organisation of their energy transfer. All are variations of the shaped charge (HEAT) in that they use a liner, but in this case a dished plate of various contours depending on the exact form of attack. The contours and material of the plate when matched to the HE charge will determine the profile of the metallic slug that is formed after initiation of the explosive. The essential difference between the various plate charges themselves and the shaped charge (see Fig. 25) is that as the angle at the "apex" of the liner (plate) becomes more obtuse, so the velocity of the leading part of the metallic slug that is formed decreases, and the velocity of the tail of the slug increases. Thus a slug is formed rather than a metallic jet plus a slug, as is the case with the shaped charge.

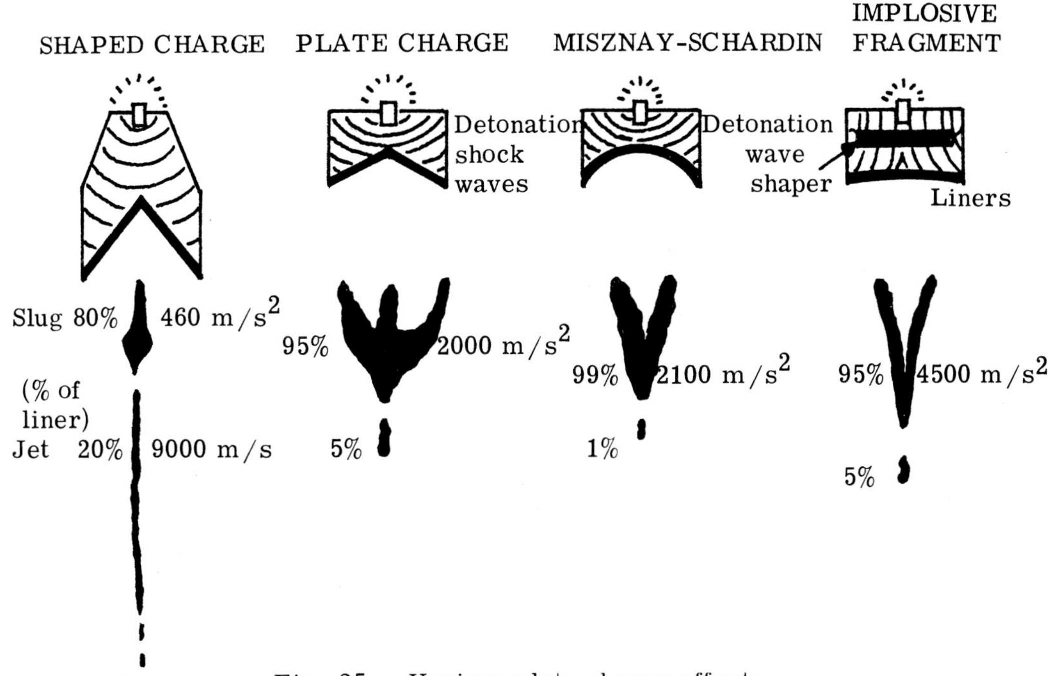

Fig. 25. Various plate charge effects

The detonating wave can be shaped to act on the plate to produce a cone shaped slug. This aids the slug's penetrative performance, because, in essence, the various slugs formed by the various plate charges are very crude kinetic energy projectiles. Penetrative performance by plate charges, therefore, is comparatively poor, particularly when compared with HEAT. If penetration is achieved, however, lethality is high, as both the attacking slug and fragments of the defeated armour contribute to the considerable behind armour damage that occurs. Furthermore, as a metallic jet is not formed there is no requirement for a "stand-off distance" to allow the jet to form and stretch. These characteristics are most effectively employed in munitions like mines which attack the more thinly armoured under-belly of a tank, or in munitions specially designed to attack APC/MICV targets or other lightly armoured and soft skinned targets (for example, aircraft).

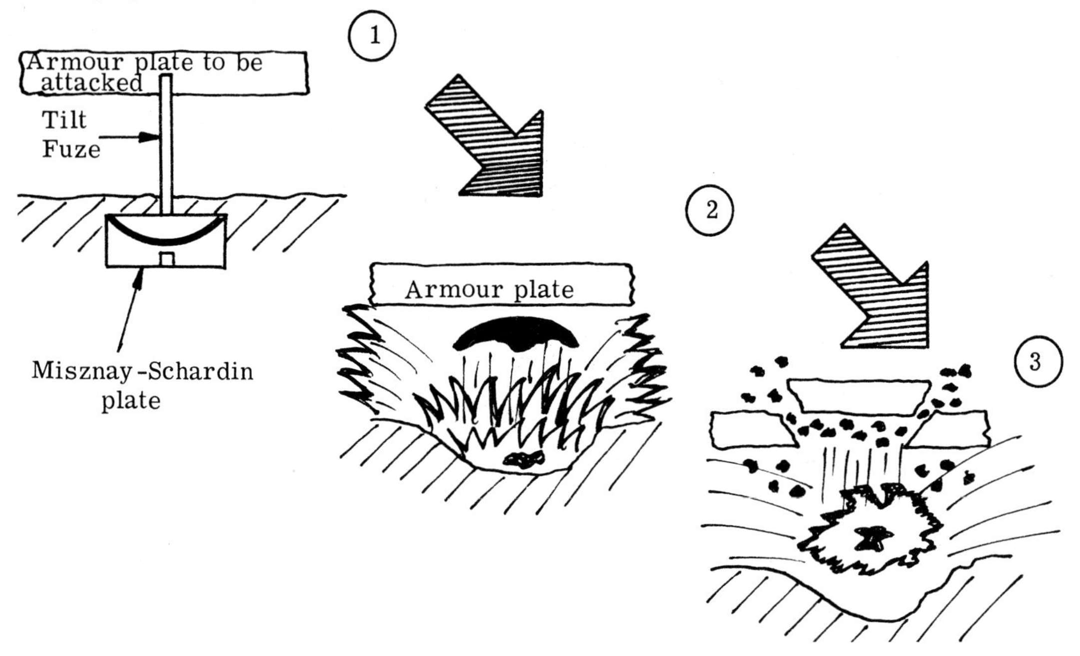

Fig. 26. The plate charge effect

Miscellaneous Anti-Armour Projectiles

There are many other types of anti-armour projectile, but the majority are variations of one of the modes of attack so far described. The armour-piercing high explosive (APHE) is one such variation, utilising both kinetic and chemical attack to produce its target effect. Figure 27 shows a 122 mm USSR APHE round.

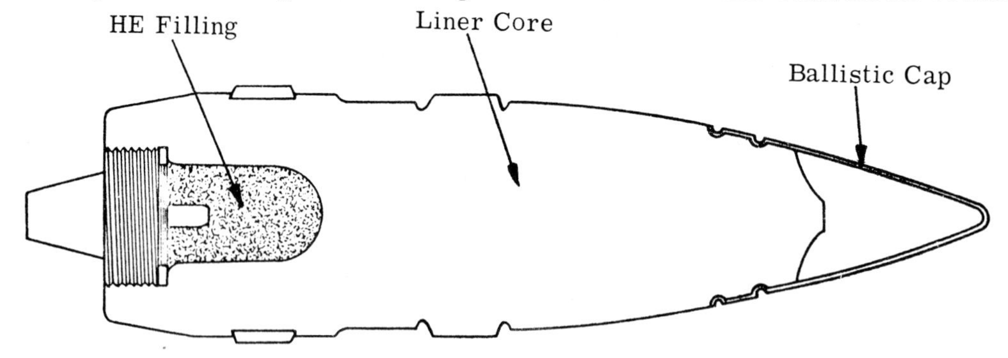

Fig. 27. USSR 122 mm APHE projectile

The round is essentially a kinetic energy solid shot, but with a base cavity filled with HE. As the AP shot penetrates the target, the HE filling is detonated and this enhances the effect behind the plate. In fact, it is not absolutely proved that the HE detonates as intended, and on balance it would seem that it detonates on strike impact (bearing in mind the velocity requirements for an AP shot). The dilemma with this nature of ammunition is that on the one hand, if the shot does not have a sufficiently high mass it will not penetrate, and by making a cavity in the projectile its mass is being decreased. On the other hand, if due regard is paid to maintaining its mass, the cavity and therefore the HE filling will be very small.

SUMMARY

There is no absolute solution as to which method of attacking armour is the best, as indeed there is no absolute solution to armour protection. All the conventional methods of attacking armour are compromises and subject to constraints which dictate solutions to the problem which are less than ideal.

A kinetic energy projectile requires a heavy and bulky delivery means. The projectile itself must be capable of withstanding the high pressures required to give the necessary high muzzle velocity. The chance of a hit is proportional to the range and the speed of the target, as there is no way of altering the flight path of the projectile once it has been fired.

HEAT projectiles require a much lower muzzle velocity than kinetic energy projectiles and should, ideally, be free from spin if their performance is not to be degraded. They can be launched by rockets or by guns, either smooth bore or if rifled by fitting slipping driving bands. A HEAT projectile is effective against spaced armour, its performance does not fall off with range, and it requires a relatively small quantity of HE to achieve very good penetration.

HESH projectiles need a low strike velocity to avoid reverse impact detonation degrading their effect. They are completely defeated by spaced armour, and require a sizeable quantity of HE to give a worthwhile effect. They are, however, very effective against a range of other targets, particularly concrete structures.

The various types of plate charges, although lacking in penetrative performance compared to the other modes of attack, are extremely lethal if penetration is achieved. They can also be initiated some distance from the target and still achieve significant penetration against light thin armours. They are an ideal means of attack for such munitions as mines, minelets ejected from a gun fired carrier shell, aircraft bombs and rockets.

By maintaining this range of attack options, the ammunition designer does force the tank designer to compromise on the degree of protection he can give the tank against any one form of attack, so as to cover the range of possible attacks. This is an important reason for keeping the range of attack options as wide as possible.

SELF TEST QUESTIONS

QUESTION 1 Describe the tank damage assessment criteria.

Answer ....

....

....

....

QUESTION 2 Draw a diagram to show the angle of attack.

Answer

....

....

....

....

QUESTION 3 Describe briefly why the tank represents such a formidable target to the ammunition designer.

Answer ....

....

....

....

QUESTION 4 Describe in outline the kinetic energy attack of armour.

Answer ....

....

....

....

QUESTION 5 What is the fundamental armour equation?

Answer ....

....

. . . .

. . . .

QUESTION 6 Describe the function of a swivel nose cap when fitted to a kinetic energy shot.

Answer . . . .

. . . .

. . . .

. . . .

QUESTION 7 Describe the penetrative paths of AP shot when it:

a. Strikes at the normal.

Answer . . . .

. . . .

. . . .

b. Strikes at high angles of attack.

Answer . . . .

. . . .

. . . .

QUESTION 8 Describe the ways in which shot can fail when striking armour plate.

Answer . . . .

. . . .

. . . .

. . . .

QUESTION 9 What are the requirements for AP shot in terms of mass (m), velocity (v) and diameter (d):

a. At the target?

Answer . . . .

. . . .

b. In flight?

Answer . . . .

. . . .

c. In the gun?

Answer . . . .

. . . .

QUESTION 10 Describe briefly what is meant by the abbreviation APCBC.

Answer . . . .

. . . .

. . . .

. . . .

QUESTION 11 What are the advantages and disadvantages of APFSDS compared with APDS?

Answer . . . .

. . . .

. . . .

. . . .

QUESTION 12 Explain briefly the hollow charge effect.

Answer . . . .

. . . .

. . . .

. . . .

QUESTION 13 Comment on the lethality/penetration relationship of HEAT ammunition.

Answer . . . .

. . . .

. . . .

. . . .

QUESTION 14 Comment briefly on how each of the following factors affect the performance of HEAT.

a. Cone (liner) diameter.

Answer . . . .

. . . .

b. Spin.

Answer . . . .

. . . .

c. Stand-off distance.

Answer . . . .

. . . .

d. Symmetry of charge and liner.

Answer . . . .

. . . .

QUESTION 15 Describe the HESH method of attacking armour.

Answer . . . .

. . . .

. . . .

. . . .

QUESTION 16 What is the most serious limitation of the HESH method of attacking armour?

Answer . . . .

. . . .

. . . .

. . . .

QUESTION 17 What other forms of attacking armour are used?

Answer ....

....

....

....

QUESTION 18 What is the essential difference between the HEAT and the various plate charge methods of attack?

Answer ....

....

....

....

QUESTION 19 Describe the main characteristics of the plate charge effect.

Answer ....

....

....

....

QUESTION 20 Describe the APHE shot and its target effect.

Answer ....

....

....

....

ANSWERS ON PAGE 253

9
The Attack of Air Targets

INTRODUCTION

In this chapter the general problems of attacking air targets are considered and details of some of the modes of attack are discussed.

TARGET ASSESSMENT

The Target

Modes of attack depend on the targets to be considered and these include: fast moving aircraft such as tactical strike, fighter bombers, fast reconnaissance types, certain missiles, drones and unmanned aircraft; slow moving types such as transport aircraft which operate at speeds between 200 and 300 knots and finally helicopters. The essential requirement is to destroy or disable the target. Any damage which prevents the target from completing its mission can be classed as disablement. In general this can only be achieved by physical disruptive effect against the main structure, the crew or the payload, but a level of destruction can often be achieved because of the dependence of this type of target on its environment.

Damage Criteria

There are various views on which method to use for assessing target damage and there is not the same agreement in this area as there is in the assessment of tank damage but the following damage levels are generally accepted:

> F_t The target will become permanently incapable of maintaining directed flight within time 't' of sustaining the damaging hit. The target fails to complete its mission and its destruction is implied.

C_t The target is unable to continue with its stated mission within the time 't' of the hit. The mission is frustrated and the target may be destroyed.

E_t Sufficient damage is inflicted on the target to cause it to be grounded for repairs for time 't' before it can undertake further missions.

As with the criteria for incapacitation of personnel there is a time element depending on the effectiveness of the mode of attack. If the warhead scores a direct hit it could produce an F kill in seconds whereas if it had a near miss the result might only be an E kill, though this could keep the aircraft grounded for several days.

Vulnerability of Target

In the same way that human targets are assessed in an elemental system so air targets are similarly treated. Each element has different characteristics and may be susceptible to different effects. Engines and the basic airframe can be quite tough as a target whereas fuel lines and distribution lines plus the crew are the most sensitive elements. Other precise elements are flight controls, avionics or 'black boxes' and power transmission systems. Orientation of target to attack has a significant influence on the result of an attack and more so on the hit probability (Fig. 1). Targets are still vulnerable in certain areas and there have been reports of modern aircraft being brought down by rifle fire in recent years.

Each vulnerable area in the target is assessed separately for each damage criterion and by relating each to the total presented area of the target the chance of a kill, give a random hit, can be determined. Clearly the aircraft designer is concerned with reducing vulnerable areas to a minimum.

Reducing Vulnerability

Aircraft are among the most complex of air targets and the reduction of vulnerability to attack is not an easy task. Nevertheless the following basic steps can be taken to reduce vulnerability: add armour protection to particularly sensitive points, bury sensitive components well inside the target and concentrate a number of sensitive components together. Duplication is also a possible solution. The addition of armour naturally brings a weight penalty and this must be balanced by the resultant protection achieved. It is not difficult to bury or hide sensitive components within or behind main structures rather than leaving them exposed in the 'shop window'. The concentration of several components does not solve the problem but it does reduce the chance of a hit causing serious damage albeit at the risk of making the results of a hit, when it does occur, more serious. Duplication requires more space and weight because the idea is to have two sets of controls or other systems if it is necessary to separate vital elements.

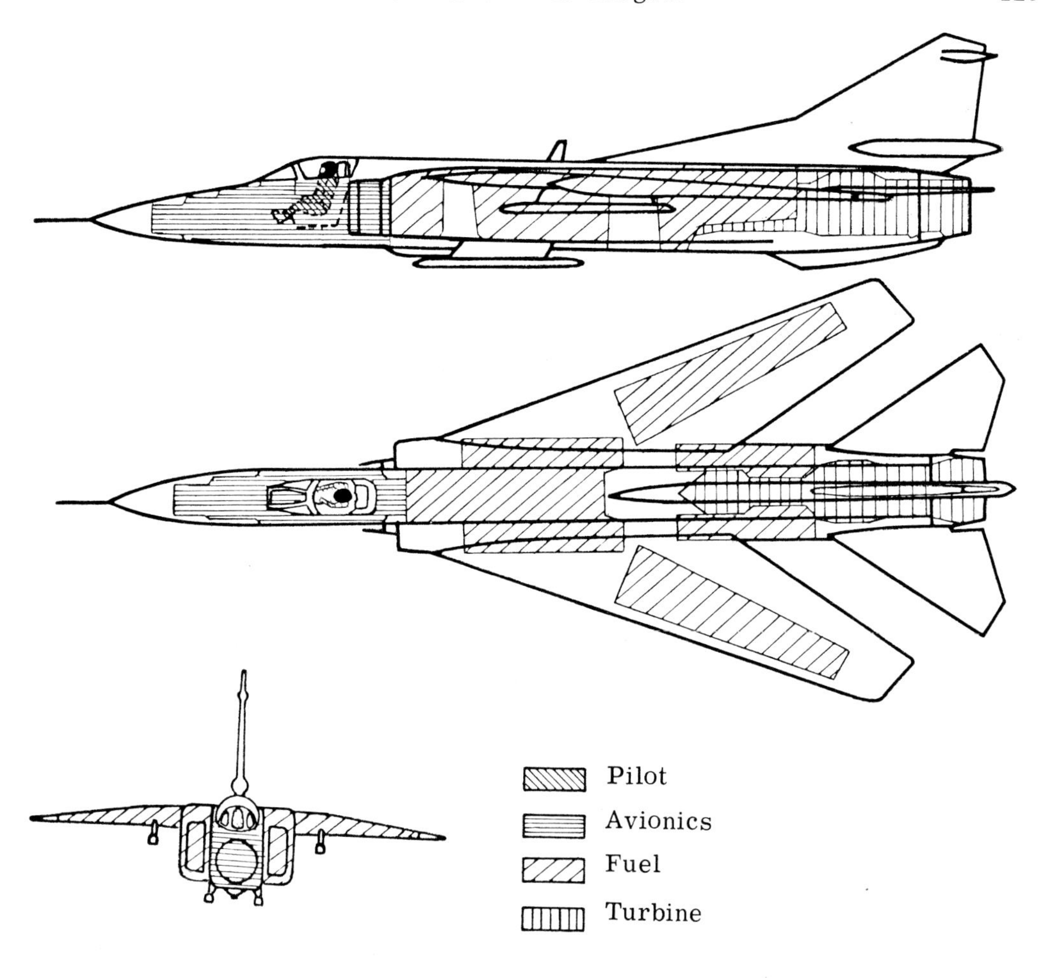

Fig. 1. Vulnerability model

Modes of Attack

Anti aircraft guns have traditionally fired high explosive (HE) projectiles at air targets. These projectiles are fitted with self destruct devices to prevent them returning to the ground in the event of a target miss which frequently occurs. During World War II an average of 187,000 rounds were fired into the air for every aircraft destroyed and against the VI Flying Bomb (350 mph) 156 rounds were fired for each casualty. There was also a gradual increase in the rate and size of projectiles fired in a search for higher ceilings and greater damage potential as a result of the target becoming much improved in performance. The calibre reached 5.25 inches firing an 80 lb projectile but this, like the others, had a relatively small high explosive filling due to the projectile wall thickness and strength required at the gun where launch conditions were very severe. Rockets were also fired to leave, in some cases, a mine in the air suspended on a wire and parachute for the target to fly into. As the target became faster, more sophisticated and perhaps tougher, the designer had to rethink the problem as conventional projectiles could not reach the higher target nor catch the faster ones. A mach 1 aircraft will travel several kilometres in the time taken for a conventional projectile to reach say 15,000 m. Clearly it became almost pointless to fire conventional guns at that particular type of target. The best that one could hope for was to put up a blanket or wall of shrapnel by mass firing in the hope that the target would be damaged. Conventional guns of calibre 40 mm and below however still play an important part in attacking air targets particularly those flying at low altitudes.

The introduction of guided missiles post World War II solved certain problems which the conventional gun designer found insurmountable. The first was the freedom to try new types of warhead and configurations now possible because of the low launch acceleration of missiles compared to gun projectiles. The second was the ability to control the missile and steer it to the target which is not possible with a conventional projectile once it has been fired from the gun. The third relates to the need and potential for increasing the size of warhead and thus its terminal effects.

Whilst the term missile is now a generally well known and accepted term, is it possible that it is designed to miss rather than hit the target hence its name? It is a very expensive process to design a missile with a high hit probability and it is often cheaper to use a larger warhead with a bigger punch and allow a miss-distance for it to operate in. This is the current trend and is usually referred to as warhead matching. There is little point in producing a system capable of a miss-distance of 1 metre if the warhead has an effective lethal radius of 5 metres.

BASIC WARHEAD TYPES

There are several types of warhead in service use for the attack of air targets and detailed information is provided in Chapter 7. Certain basic characteristics however will be discussed here to complete the overall picture.

Blast

Blast is produced by a bare charge of high explosive but for obvious reasons there is a need to contain the explosive in a case of some sort. To be effective a blast wave needs a medium through which it can travel and as the altitude of operation increases air becomes less dense thus reducing the blast effect. At 30,000 m about twice as much explosive is needed to give the same effect as at sea level. The casing is usually light and relatively thin and provides very little fragmentation effect. The miss-distance for blast type warheads is clearly rather small particularly at high altitudes and the charge to weight ratio is about 5:1. Both internal and external blast warheads may be found, the difference being one of target effects. Internal blast types are designed to enter the target and detonate inside whilst the external warhead is designed to actuate outside and near to the target. The blast warhead achieves its target effects purely by an over-pressure commencing with a positive impulse phase followed by a comparatively short negative impulse as shown in Fig. 2.

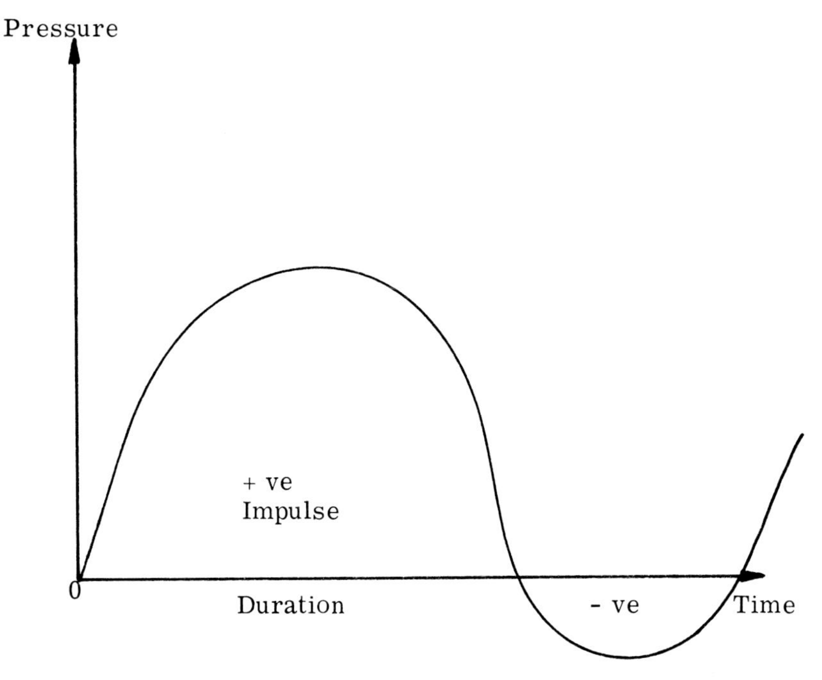

Fig. 2. Blast profile

Blast effects fall off rapidly with distance and the following table, Fig. 3, shows the weight of explosive required to give the same damage effect at various distances from point of burst.

DISTANCE FROM POINT OF BURST	WEIGHT OF EXPLOSIVE TO GIVE CONSTANT DAMAGE
10 m	10 kg
20 m	80 kg
30 m	270 kg

Fig. 3. Distance and weight of explosive relationship

Although these relationships are not accurate it will be seen that blast warheads need to be quite large if they are to be effective even at miss-distances of a few metres at high altitudes.

Fragmentation

Because of the restriction on blast effects the next to be considered is the fragmentation type in which fragments are projected to the target by the detonation of the contained high explosive. This warhead is the natural follow on from the conventional shell and indeed originally took the same shape and profile as a traditional shell or bomb being mounted in a frame. With low launch acceleration much more can be made of the terminal effects because the arrangement of metal around a charge of high explosive can be achieved in a variety of ways. Designers of warheads moved away from natural fragmentation, where a metal case was fragmented by the detonation of the high explosive inside it, to much more cost effective types. Among these are the pre-formed or controlled fragment warheads where the metal is prenotched or assembled as separate fragments into a resin bond and encased in a thin container. Another system is to surround the high explosive with long rods either mounted separately or welded together. This type was introduced because it was thought that, although larger fragments were needed to penetrate air targets than personnel, rods hitting side-on would inflict more damage. However such rods when separated are not aerodynamically stable and they tumble in flight so that a number of impacts will be end-on. In general this type of warhead is of limited use but the welded rod type has had more success. Other types incorporate spheres, cubes and various aerodynamic shapes for fragments.

Fragment Patterns

Ideally all fragments should be directed towards the target. This can be partially achieved by arranging for the fragments to be mounted inside a specific beam width. The beam width can be controlled to some extent by the external shape of the warhead (Fig. 4).

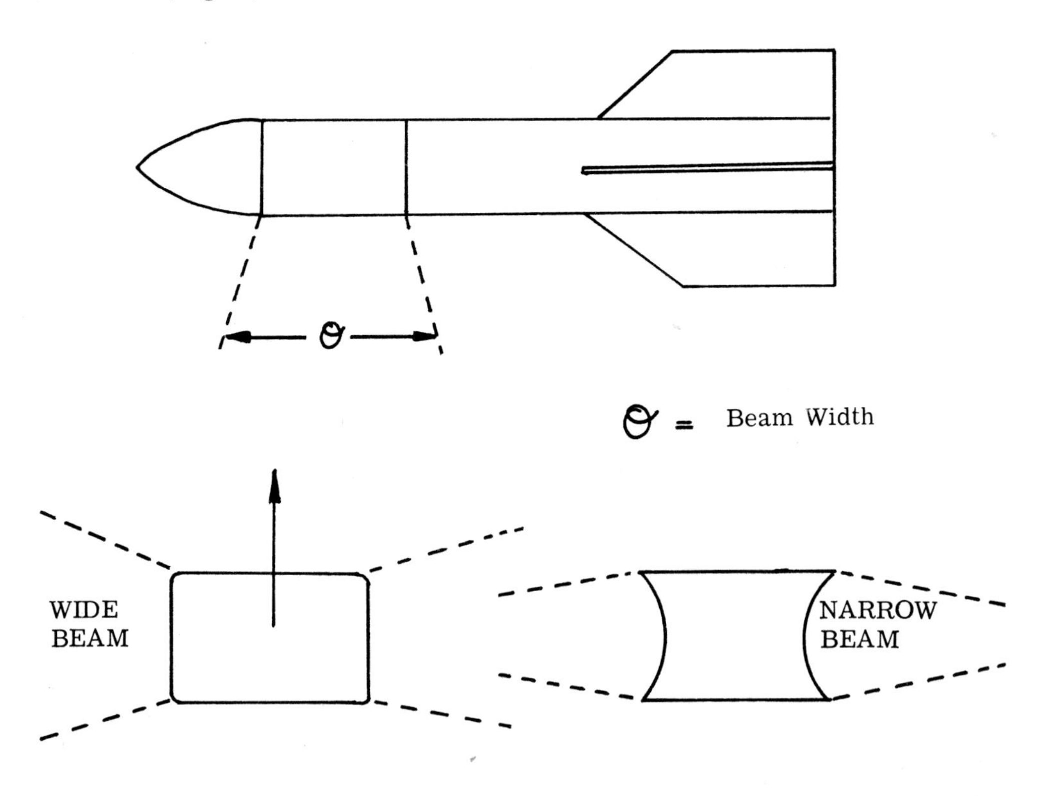

Fig. 4. Beam width

The arrangement of beam width must of course be integrated with the approach angle of the missile to the target.

Shaped Charge

Details of this type of attack appear elsewhere in this book but it is necessary here to compare the specific requirements of air target attack with that of armour targets, Fig. 5.

	ARMOUR TARGETS	AIR TARGETS
Cone Angle	40°-60°	90°+
Cone Diameter	125 mm	200 mm
Cone Material	Copper	Aluminium
Liner Thickness	2 mm	12 mm
Type of Jet	Short and thin	Long and fat

Fig. 5. Comparison of shaped charge requirements

The reasons for the different values shown in the above table are because a long fat jet is required with a long travel (due to miss-distance) and made up from aluminium to increase the incendiary effect in the air target. Generally low density liners are used for low density targets and high density liners are used for armour. The stand off distance for armour attack is of the order of 4 calibres but for air targets, again because of miss-distance, stand off is very much more.

Other Types

There are various other types of warhead, some of which are mentioned in Chapter 7 (Warheads).

Position of Warhead

Unlike conventional weapons, missile warheads are not and cannot always be placed at the front end of the missile due to radome, sensor and other requirements. This, to a certain extent influences warhead design, in particular shape and weight. It is perhaps worth considering whether, if a missile were to make direct contact with an air target, a warhead would be needed?

CONCLUSION

Current thought on attack of air targets is to use a mix of high rate smaller conventional guns for low flying targets and guided missiles for all types of target. It is difficult and costly to design a warhead with a high probability of a target kill unless the miss-distance is small and the intercept profile consistent. Considerable work is going on in an effort to increase warhead lethality but a more profitable approach may well be to reduce the missile distance and use a relatively simple warhead.

SELF TEST QUESTIONS

QUESTION 1 What are the generally accepted criteria for air targets?

Answer ....

QUESTION 2 How are air targets assessed?

Answer ....

....

....

QUESTION 3 Explain how vulnerable areas can be reduced in air target design.

Answer ....

....

QUESTION 4 Why are missiles used against modern air targets in preference to traditional anti-aircraft guns?

Answer ....

....

QUESTION 5 List the current basic types of warhead used against air targets.

Answer ....

....

QUESTION 6 Compare the shaped charge used against air targets with the type used against armour.

Answer ....

....

....

QUESTION 7 Why are there different requirements for shaped charges?

Answer ....

....

QUESTION 8 Why is it necessary to have a warhead in a missile?

Answer ....

. . . .

. . . .

QUESTION 9 Explain the importance of matching the warhead to the overall missile system.

Answer . . . .

. . . .

. . . .

QUESTION 10 Why is blast less effective at high altitudes?

Answer . . . .

. . . .

ANSWERS ON PAGE 256

10
Carrier Projectiles

INTRODUCTION

Most weapon systems have a variety of projectiles which can be fired for specific purposes. High explosive projectiles have been dealt with in Chapter 6 therefore this chapter will cover the rest,which are known collectively as carrier types. There are many but they all operate in one of four ways which are described below. Generally speaking a carrier projectile is one which relies on its payload to give the desired effect at the target, the projectile body being used merely as a carrier.

HISTORY

The spherical shell of 1804 was perhaps the first carrier type to be used and it had a tendency to 'explode' prematurely, possibly because the shrapnel and powder were mixed together and produced a reaction on firing. In 1854, the shrapnel or metal balls (bullets) were separated by means of a diaphram at the suggestion of Colonel Boxer which reduced the problem of prematures. Later the sphere gave way to the elongated shell, a shape which is more in line with current projectiles. Fig. 1 shows an example of both the spherical and elongated shapes.

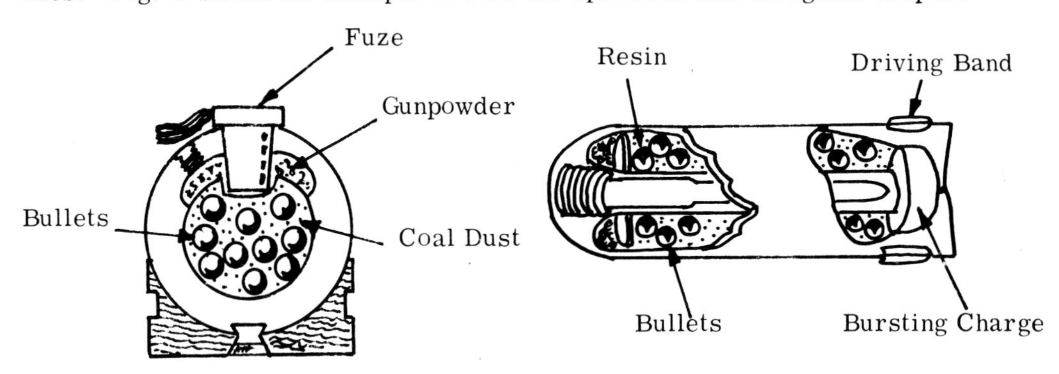

Fig. 1. Early shell

Shrapnel was and is used against personnel but it has also been used for wire cutting and in the anti-aircraft role. Projectiles were made from cast iron initially but later, much later, steel was used. There were other types of carrier projectile used before the modern types were designed such as case shot which was an anti-personnel device consisting of a thin metal case containing round bullets packed in clay or sand. Examples are shown in Fig. 2 and it is interesting to note that the modern version is called canister.

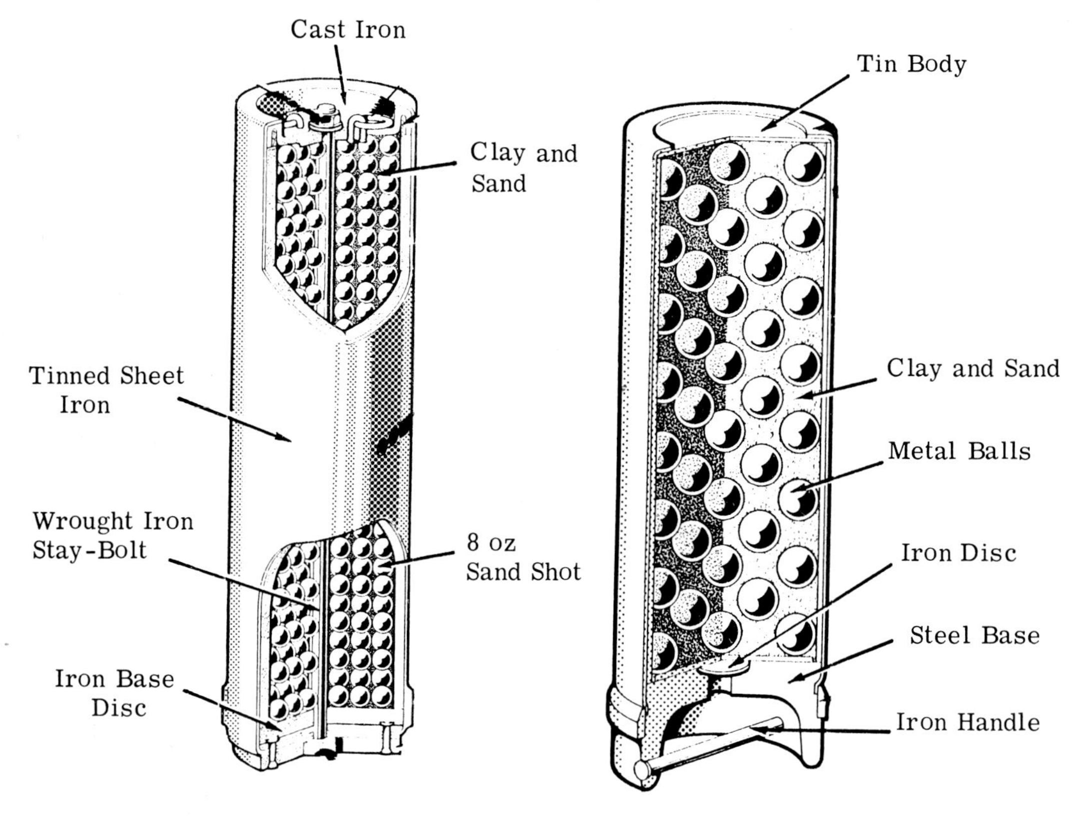

Fig. 2. Case shot

METHODS OF OPERATION

There are four basic methods for the operation of carrier projectiles. These are bursting, base ejection, nose ejection and base emission.

Bursting

This type of projectile consists usually of the normal high explosive projectile body manufactured in the same way and comprising a similar profile, but filled with substances other than high explosive. It is actuated at the target by impact and ideally should range the same as the high explosive projectile. It is fitted with a disruptive fuze which, in conjunction with a small quantity of explosive, opens the projectile to scatter its contents on impact. The basic design is shown in Fig. 3. There are, however, variations on the design which will become apparent later.

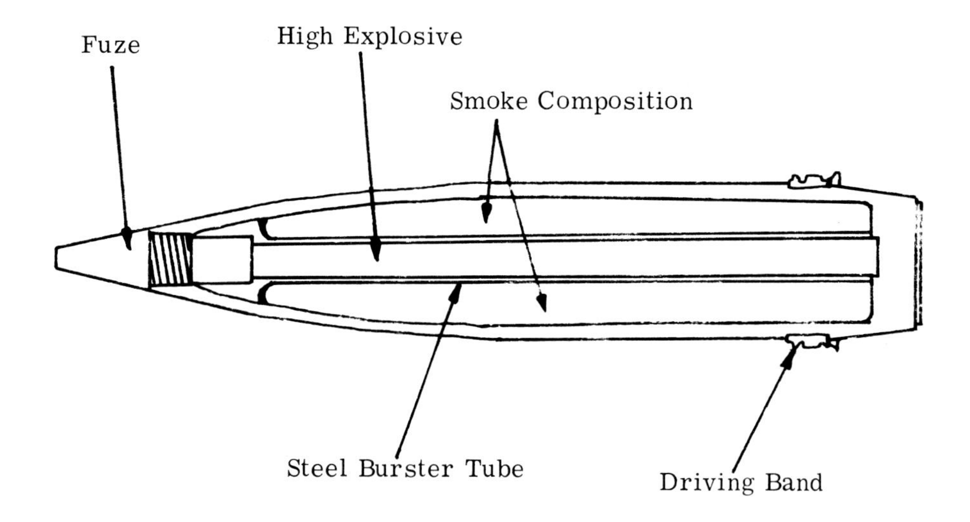

Fig. 3. Bursting type carrier shell

Base Ejection

This is the most common design and, as the name implies, is operated on the base ejection principle where the payload is ejected through the base of the projectile usually during flight. Because of this requirement the projectile body is different from the normal high explosive type and the ranging requirement does not necessarily apply. It is essential for the projectile to have a cylindrical internal cavity and a weakened base to ensure efficient projection or ejection of the contents. The base is enclosed by a plate which is pinned by twist or shear pins or screw threaded by a few threads only. Other methods of attachment may be found, but all methods are designed to give strength during firing yet allow easy removal by inside forces when the projectile is needed to operate. An example of this type is shown at Fig. 4. The contents are ejected by means of a burster charge which is initiated by a time fuze and provides pressure inside the projectile to eject the contents which initially tend to follow the trajectory then disperse as appropriate. The flash from the burster also ignites the payload container if necessary.

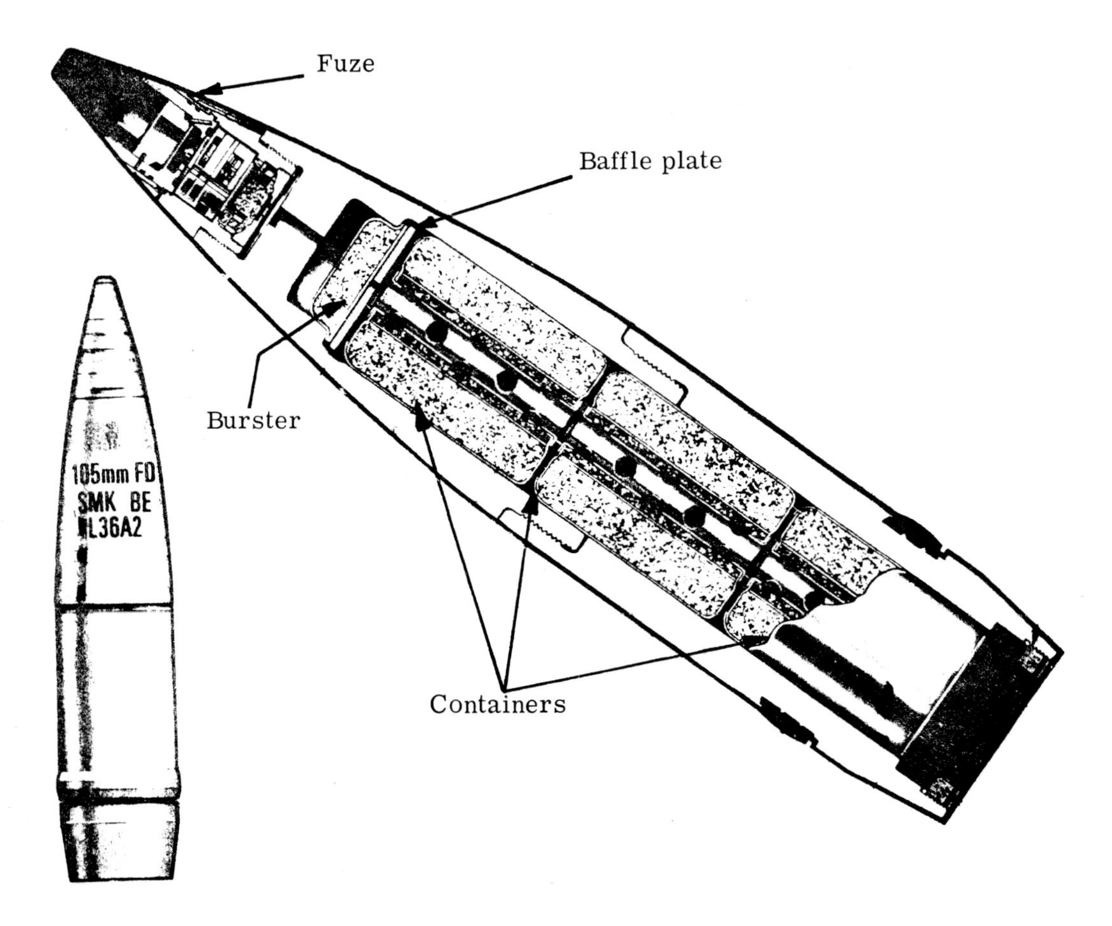

Fig. 4. Base ejection

Nose Ejection

The nose ejection projectile design is similar in operation to the base ejection type except that the payload is ejected forward. The projectile has a weakened section towards the nose at the end of the parallel position of the body. The flash output from the magazine of the time fuze passes down a central channel to ignite a burster. There are no current projectiles which use this system in its entirety but some have a modified form of combined bursting and nose ejection.

Base Emission

The base emission method of operation is generally met only in weapons such as tank guns and mortars where it has been used for smoke projectiles. The smoke mixture is ignited by propellant gas heat when fired with the priming composition and delay fitment located in the projectile base. At the target the smoke is emitted through holes in the base. An example is shown at Fig. 5.

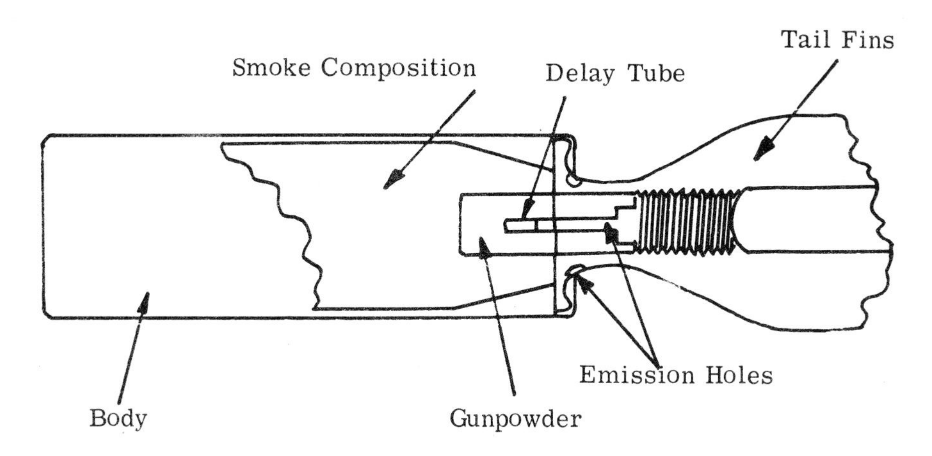

Fig. 5. Base emission

TYPES OF PROJECTILE

Bursting White Smoke

Bursting white smoke projectiles are filled with white phosphorus (WP) which ignite spontaneously when the projectiles are broken open by the small quantity of explosive present. The HE is normally filled into a central column and surrounded by the WP. Early bursting projectiles merely had a HE burster to break open the nose of the projectile. Although a typical example of a bursting WP projectile is shown in Fig. 3, Fig. 6 below shows a based fuzed bursting smoke projectile designed for a tank gun.

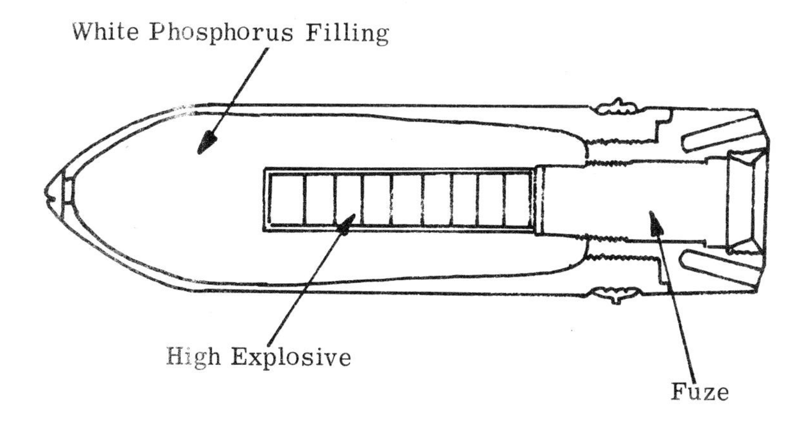

Fig. 6. Tank bursting white phosphorus smoke shell

Bursting Coloured Smoke

A bursting coloured smoke projectile is filled forward of the driving band with a pelleted mixture of PETN/Paraffin wax/dye, the dye being used to give red or orange colour effect. Behind this is a pellet of wax/dye composition. The shell is fuzed and has a high explosive pellet beneath it to break open the projectile and initiate the contents. The projectile is normally used to give a coloured burst for target indications. An example is shown at Fig. 7.

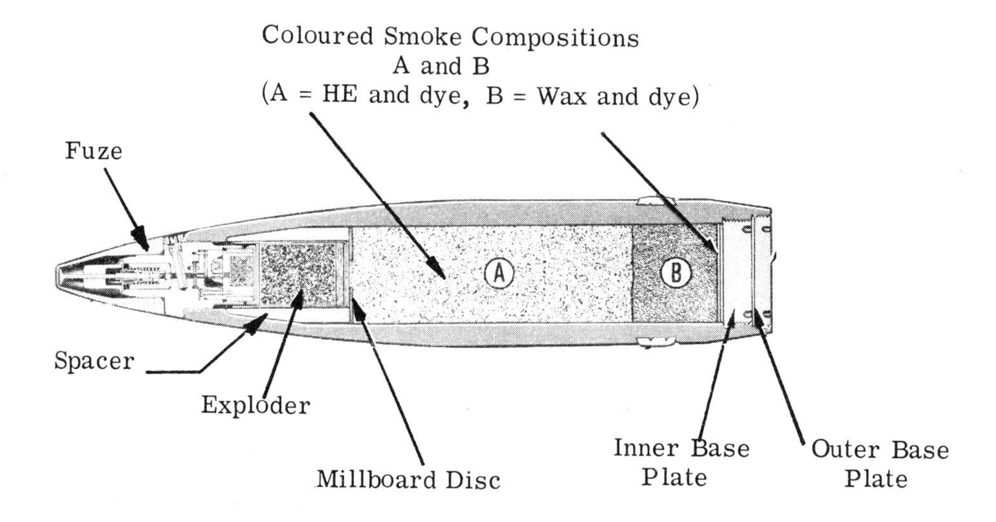

Fig. 7. Bursting coloured smoke shell

Base Ejection Smoke

The base ejection smoke projectile is fitted with a time and percussion fuze which has to be pre-set. Three or more containers filled with smoke composition are fitted inside the projectile and these are ejected during flight. The smoke composition is usually hexa-chlorethane (HCE) and, on ignition, the HCE reacts with zinc oxide giving zinc chloride which is enhanced in its smoke effect by moisture in the atmosphere. Coloured smoke is obtained by the inclusion of the appropriate dye plus sugar and potassium chlorate. Containers may be open at one end or closed at both ends with a central perforated channel to allow ignition. Length of containers must not be less than 0.8 x diameter or they may jam on ejection or topple during dispersion. Details of various containers are shown in Fig. 8. The baffle plate fitted to the projectile beneath the burster is designed to assist the build up of pressure prior to ejection, the central hole allowing the flash to reach each container. Variations in time or pressure of ejection are achieved by altering the quantity and fineness of the burster composition. The flash from the burster is picked up by the sleeve of shalloon which is primed and this is passed on to a priming composition. Priming is

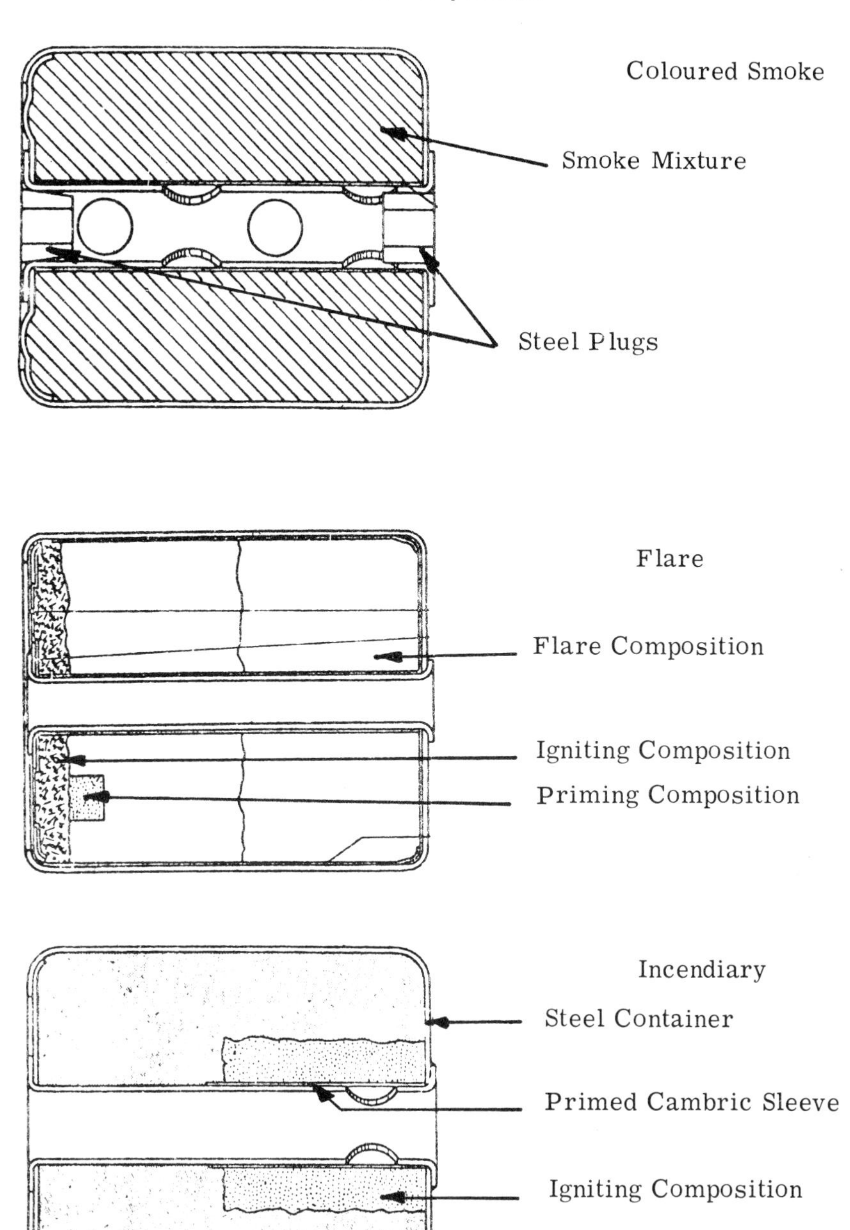

Fig. 8. Typical containers

necessary because HCE is difficult to ignite and this results in a 10-20 second delay before the full rate of smoke emission. A typical base ejection smoke projectile is shown below in Fig. 9, which also shows the operation.

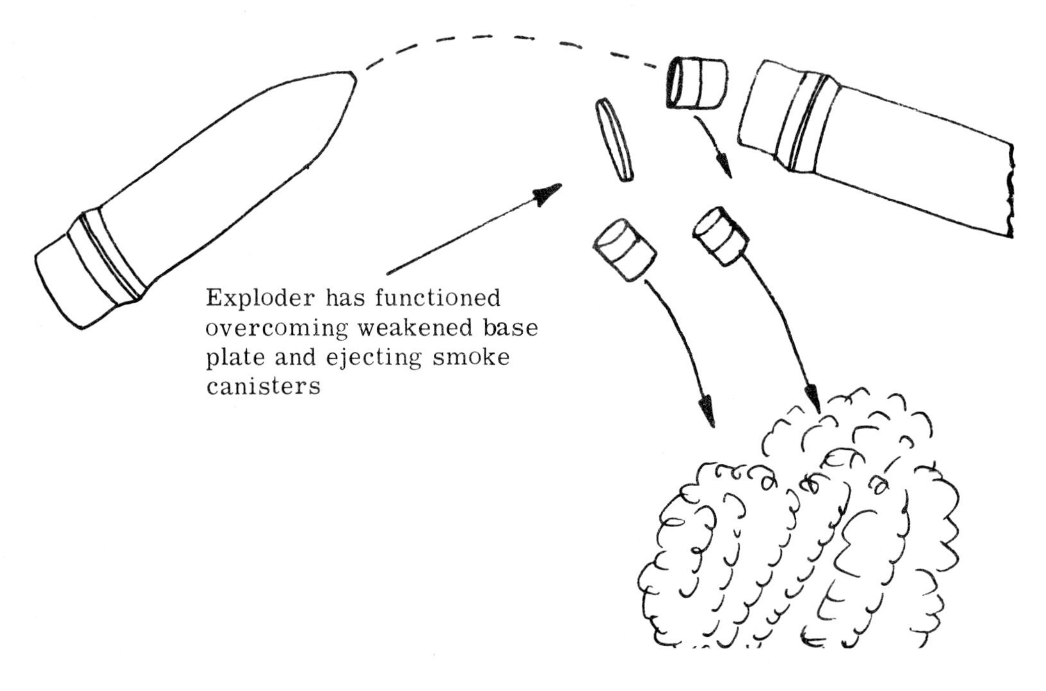

Fig. 9. Operation of base ejection system

Comparison of Bursting and Base Ejection Smoke System

A comparison of the two types of carrier smoke projectiles are shown in the table in Fig. 10. In the case of bursting types, due to the heat of reaction, there is an immediate cloud of smoke produced which is ideal for emergency use. There is a bonus incendiary effect and this does have an anti-personnel effect. Unfortunately this heat of reaction tends to produce a pillaring effect which is not advantageous except in hilly country. The white phosphorus type of filling has, in the past, been unpopular with users but with modern filling techniques and sealants the incidence of leaking is very rare. For a slower but more sustained smoke screen the base ejection type is perhaps best suited provided the containers do not become too widely dispersed. Their flat profile prevents them being buried on contact with the ground, except in soft ground, snow etc, but they may be subject to a bounce effect on rocky or very hard surfaces.

Base Ejection Illuminating

There is always a need to illuminate the target or the battlefield and this is achieved by ejecting an illuminating candle suspended on a parachute. The

COMPARISON OF BURSTING (W. P.) AND BASE-EJECTION (H. C. E.) SMOKE PROJECTILES

	BURSTING (W. P.)	BASE-EJECTION (H. C. E.)
1. Smoke Produced quickly.	YES	NO
2. Duration of smoke.	NO	YES
3. Pillaring.	YES	NO
4. Effects in soft ground.	Buries therefore reduced effect.	Buries in deep snow.
5. Percussion Fuzing (To save inaccuracy and setting) (delay of time fuzes.)	YES	NO
6. Ranges with H. E.	YES	Difficult if H. E. is S/L.
7. Smoke produced where projectile lands.	YES	Containers bounce on hard or hilly ground.
8. Secondary effects.	YES incendiary	NO
9. Independent of atmospheric humidity.	YES	NO
10. Economical Production.	More Expensive	YES
11. Safety in Storage.	Fire hazard when leaking	YES

Fig. 10. Comparison of bursting and base-ejection smoke projectiles

parachute is assembled at the base end of the projectile away from the burster and is usually protected from being damaged by the set back force on firing by 2 steel semi-cylinders. The steel semi-cylinders also provide internal rigidity and assist the illuminating composition to take up the spin of the projectile. The parachute is fireproofed and the ball joint between the shroud lines and the illuminant obviates entanglement or twisting of the lines. Burning times of these carrier projectiles can be of the order of 1 minute giving a very bright light. An example is shown at Fig. 11 which also shows the operation.

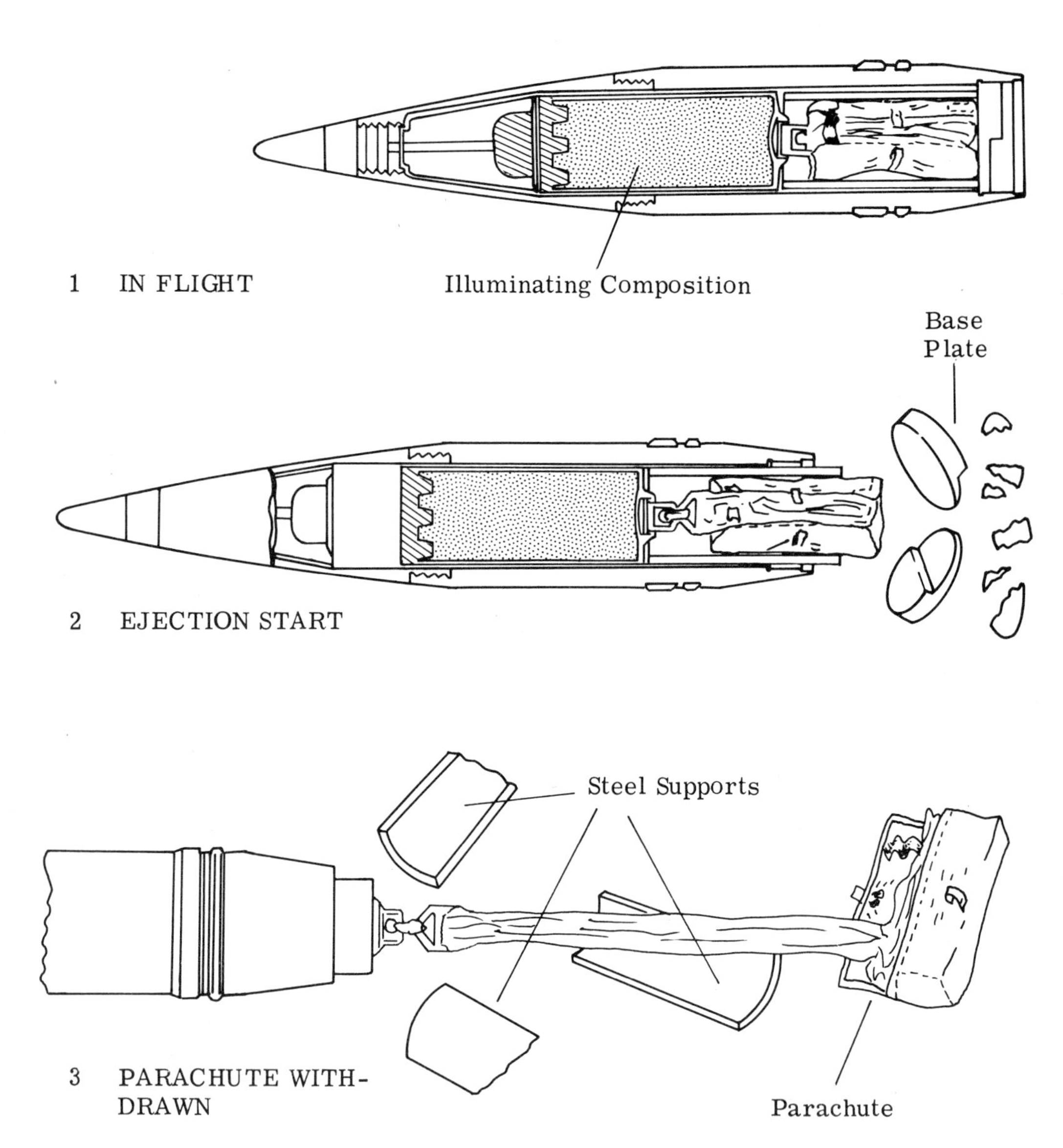

Fig. 11. Base-ejection illuminating projectile

Other Base Ejection Projectiles

There are other types of base ejection projectiles which operate in the normal way. These include propaganda, radar echo, incendiary and flare payloads. The propaganda type is filled with small leaflets which are dispensed in the air giving random distribution on the ground. Radar echo (RE) projectiles consist of thousands of aluminium or foil spills or brads which hang in the air to give a radar response. A flash pellet may be included for visual observation. RE is used to confuse enemy radar and also to provide a ranging mark for meteorological purposes for use by friendly radar. Incendiary payloads are housed in steel or magnesium alloy containers and provide incendiary effect at the target. Flare fillings are used for night signalling and target indication. There are other types of base ejection carrier projectiles but they generally conform to the standard design.

Anti-Personnel Carrier Projectiles

To complete the chapter, mention must be made of some anti-personnel type carrier projectiles. It was mentioned earlier in this chapter that the first carrier projectile was the shrapnel type and this was possibly the first attempt at controlling fragment size on a bursting projectile. A similar approach is met in a number of current anti-personnel mines and in one imported weapon using a high explosive projectile. Although there is no current carrier projectile using this system there is one which adopts another very early principle as suggested earlier and this is the canister projectile. This is based on the old case shot idea and was developed to counter the threat posed by massed infantry to unsupported armoured vehicles. It is essentially a short range device - up to about 300 metres - and consists of a thin metal case containing, in a tightly packed formation, a number of similar steel fragments either spherical or cylindrical in shape. The container breaks up at the muzzle of the gun and the contents are dispersed forwards in a dense conical pattern. An example is shown in Fig. 12.

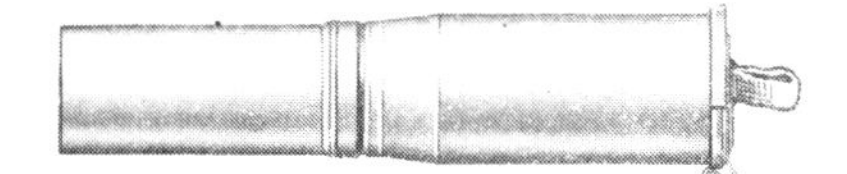

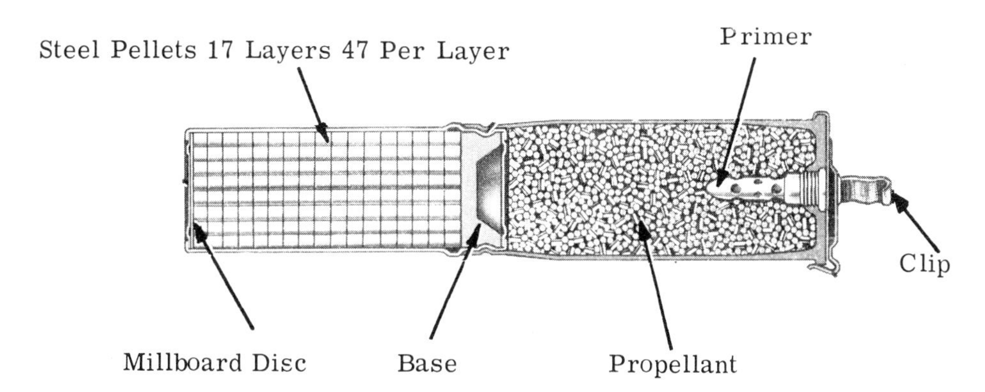

Fig. 12. Typical canister round

Finally, a more recent anti-personnel type known as the "beehive" or flechette projectile consists of a payload of small dart like shapes which are projected forwards and outwards from the projectile. The projectile is fuzed so that it can be used in the direct fire role as for canister or as an airburst system.

Future Trends

Large calibre weapons are now being used to convey bomblets and minelets of various configurations in a projectile body for use against various targets such as personnel, armour, motor transport and fortifications. These tend to be very expensive and are therefore selective in their use. Some design concepts are shown in Fig. 13 and they could contain such shapes as spheres, cubes, near cubes, glider and wedge.

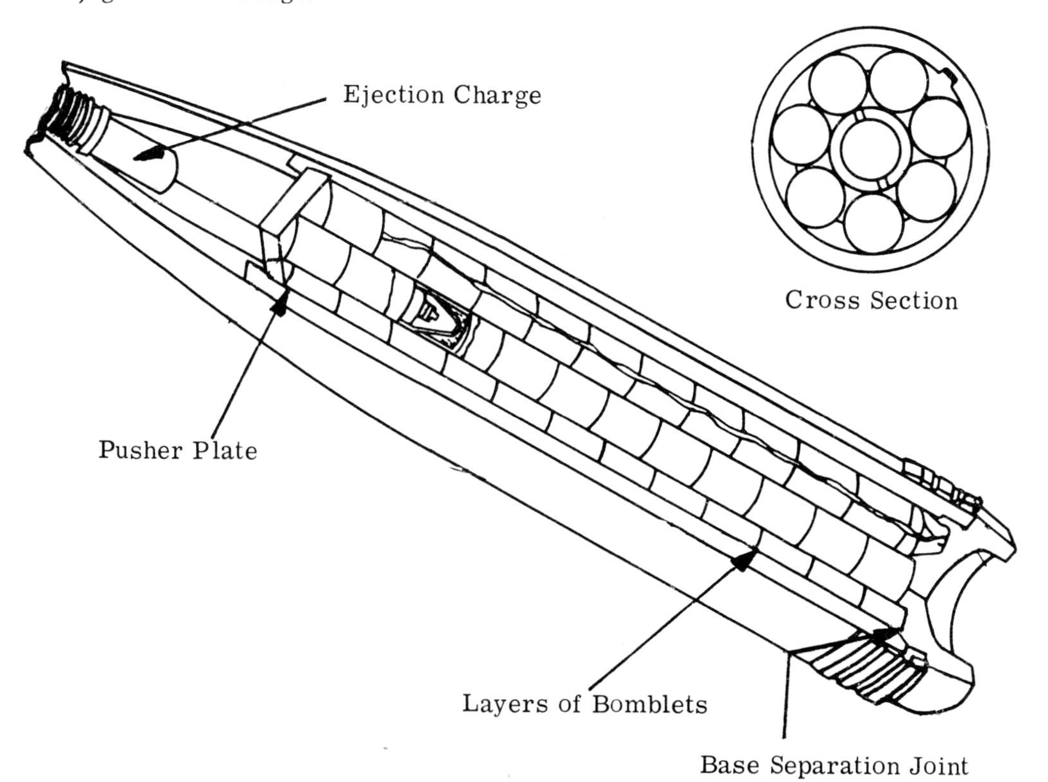

Fig. 13. Bomblet and minelet shell

Variations

There are other projectiles which are not necessarily carrier types which may be encountered and brief notes are recorded below. A target shell, usually a white phosphorus bursting projectile, is used as an anti-aircraft target by actuating it high on its trajectory.

Various types of practice projectiles are used for training in accuracy and spotting. These are usually high explosive bodies filled with high explosive substitute (HES). Special empty projectiles may also be approved for these tasks. Break up shot are used for testing mechanisms in weapons and for training in loading and firing without the need for a special firing range. They are made from steel or bakelite, weighted to compare with service projectiles and they break up outside the muzzle.

Proof shot are used for the proof of guns, carriages and propellant and fired into butts. They have a flat head so that they lose velocity rapidly and can be recovered. Due allowance is made when calculating the external ballistics of such shot. Another type of proof shot is for proof of fuzes and this is basically a high explosive projectile filled HES to the correct weight. It has a small flash filling to show point of burst and is designed to give similar accelerations and impulses as for service high explosive shell. An unusual projectile made from thick cylinders of paper and filled with steel dust shot which breaks up at the muzzle is the paper shot. This is used to operate recoil mechanisms of guns in static mountings or in tanks. They are designed to provide the same recoil effects as service projectiles and the length and total weight is increased to compensate for lower density and any loss of recoil due to escape of gas. Finally, the water shot, as for paper shot but filled with water on the gun positions, is used in a similar way but, due to its excessive length, it may be supplied in two or three portions for ease of handling.

CONCLUSION

The wide variety of carrier projectiles provide the services with most target effect requirements. There is still scope for more refinements, longer burning times, greater luminosity as new chemical combinations are discovered. This is an important area for future development.

SELF TEST QUESTIONS

QUESTION 1 What are the 4 methods of operation for carrier projectiles?

Answer ....

....

....

....

QUESTION 2 Explain the differences in operation between a white phosphorus filled and a red phosphorus filled projectile.

Answer ....

....

QUESTION 3 Give a brief description of the base ejection system.

Answer ....

....

....

QUESTION 4 What methods of igniting the containers are available?

Answer ....

....

QUESTION 5 Compare bursting and base ejection type smoke projectiles.

Answer ....

....

....

....

....

....

....

....

....

....

....

QUESTION 6 How is the parachute protected in an illuminating projectile?

Answer ....

QUESTION 7 What types of anti-personnel carrier projectiles can be identified?

Answer ....

QUESTION 8 List the various possible bomblet and minelet shapes.

Answer ....

....

QUESTION 9 What is understood by HES?

Answer ....

QUESTION 10 Give a brief description of a canister projectile.

Answer ....

....

ANSWERS ON PAGE 256

11
Fuzes

INTRODUCTION

With the development of projectiles it became necessary to produce devices for activating them. Crude fuzes were employed from about the 16th century and these were filled with mealed powder, sulphur and saltpetre. They were ignited initially by hand when the gun fired and later by the propellant gases. In the 18th century a time fuze was developed consisting of a wooden plug with a cavity filled with gunpowder which could be cut at a suitable position to terminate the slow burning at a required time thus 'sparking' the bursting charge of the projectile. In the mid 19th century the Boxer fuze was introduced and this was a great improvement. It was a shaped wooden cone and fitted into a projectile rather like a cork in a bottle. Also about that time a wooden concussion fuze was invented by Freeburn and a percussion fuze by Pettman. Many fuzes have been designed, developed and used since then and many are now obsolete but they all employ certain basic devices some of which will be covered in this chapter. Fuzes will be found in most projectiles, mortar bombs and mines and in some grenades and pyrotechnics.

REQUIREMENTS OF A FUZE

A fuze, which is a device used to initiate an explosive store at the correct time and place, must ensure that the store is completely safe to store, handle and fire and then function reliably when required. The general requirements are safety and reliability, and to a lesser extent, conformity with a standard profile. They should have an appropriate multipurpose function and be easy to set in the field. It should be noted here that fuzes in guided missiles are usually electronic devices and the explosive function devices are contained in a separate Safety and Arming Mechanism.

Safety

British designed fuzes are among the safest in the world and are tested to a very high standard against premature functioning. Prematures in weapons create damage to own troops and equipment and have a serious demoralising effect. It is therefore essential that ammunition, particularly fuzes, are designed to be safe under all storage transport and firing conditions. The Ordnance Board (OB) "Design Safety Principles for Fuzing Systems" are recommended as the basic design objectives for the safety features of fuzing systems.

Reliability

Although less stringent than that of safety, reliability is still an important requirement if correct and timely functioning is to be assured. Generally, the more complex the design the less reliable it may be, therefore, an abundance of safety devices and mechanisms may reduce its reliability. Despite the need to make fuzes safe in modern high performance weapons the reliability is of a very high order. Service environment, rough usage, vibration, climate, extremes of temperature and so on are all taken into account by the designer and this awareness, coupled with progressive testing and assessment during production, result in a high quality product.

CLASSIFICATION

The methods for the classification of fuzes are; Position, Function and Filling. The position relates to the part of the parent ammunition store to which the fuze is attached. Traditionally this classification refers to projectiles only and consist of nose or base types. Nose fuzes are always fitted to HE artillery projectiles and base fuzes to high explosive squash head (HESH) projectiles. Examples are shown in Fig. 1.

The classification by function is more commonly used and refers to such actions as time, proximity, percussion graze, percussion direct action, and delay. This system identifies the type of fuze involved and its modus operandi. Finally the classification, not part of the nomenclature, but a form of technical classification is the type of filling or explosive train. There are two types related to the explosive output of the fuze namely igniferous or disruptive. Many modern fuzes are multipurpose, eg "Percussion direct action and graze", "Time and percussion" and most confirm to current NATO standard sizes and intrusions.

FORCES ACTING ON A FUZE

The forces acting on a spun projectile also act on the fitted fuzes. All components in a fuze are either fixed relative to the fuze or are free to move within certain limits. The movement of free components, controlled or restrained as necessary by friction or springs or both, depends principally upon the forces arising from the firing environment. Forces can range from spin, acceleration, deceleration and, in some cases 'side-slap' from rifled weapons to merely a pressure applied to a fuze in a mine. These forces are utilised by the designer to provide mechanical devices which fulfil the basic fuze requirements. The process of arming is the unlocking of the devices at the required time in the correct sequence.

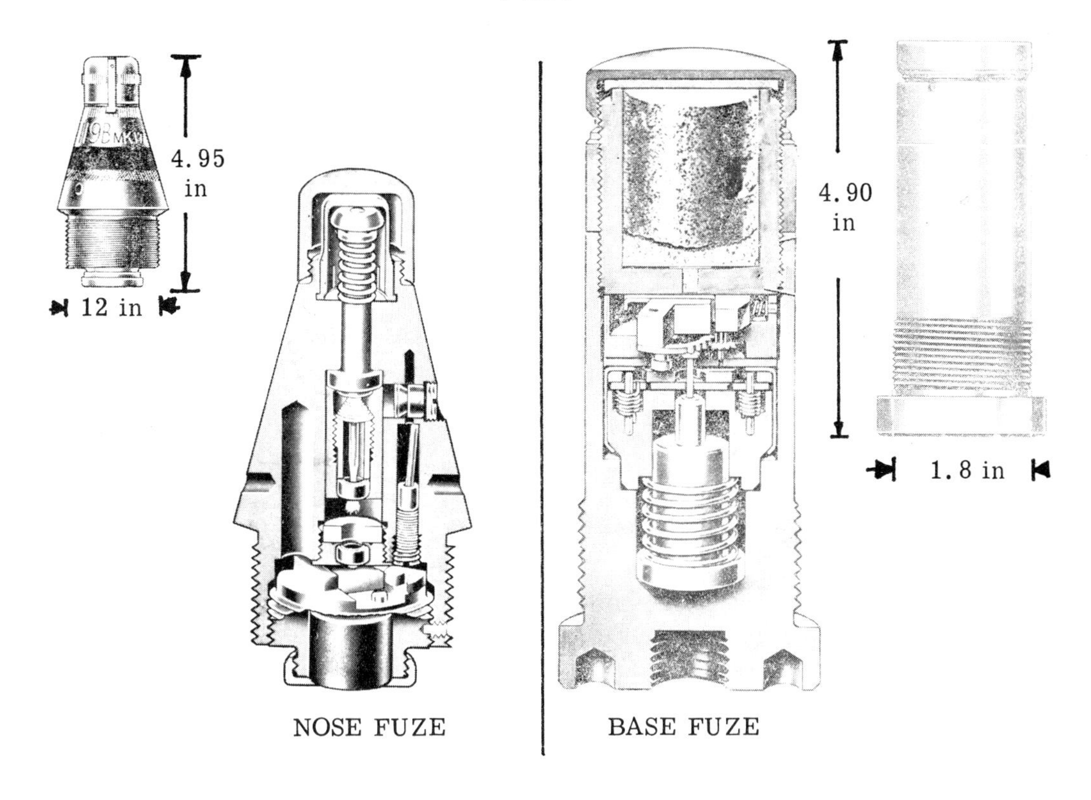

Fig. 1. Nose fuze and base fuze

Acceleration

Acceleration tends to leave components behind producing set-back on violent acceleration or creep-back on moderate acceleration.

Deceleration

Deceleration affects fuzes in two ways. After firing there is slight deceleration due to air resistance at the front and absence of propellant gas pressure at the rear, this allows a 'creep-forward' of components. On impact there is a sudden deceleration which results in 'set-forward' of components.

Spin

Centrifugal force acts at right angles to the line of flight and results in free components being flung outwards. This force cannot normally have a significant effect whilst in the barrel because the set-back forces operating are usually strong enough to create sufficient friction to prevent loose components from moving away from their set-back position.

Side-Slap

In worn guns side-slap sometimes provides a jolting effect on the fuze which could dislodge free components from their set-back position. This can be overcome by locking components 'open' and 'closed' but it is not normally a common problem.

Pressure

Pressure is normally associated with contact fuzes as used in mines and is often the only force available for the designer to utilise in static mounted ammunition.

MECHANICAL DEVICES

These devices can be dealt with under three sections namely Holding, Masking and Firing.

Holding Devices

Holding devices ensure safety at least up to leaving the weapon and provide restraints on the striker thus preventing it from initiating the explosive train. Some, such as shear wires and axial detents, are one stage devices whilst others such as centrifugal balls and arming sleeve combinations, are multi-stage types. Examples are shown in Fig. 2.

Masking Devices

Masking devices provide a positive barrier in the explosive train thus preventing prematures in the event of an initiation of the detonator during firing. There are two basic types, delay and non delay and these are usually known as shutters or interrupters. Non delay arming or masking shutters consist of sliding or rotating blocks of metal which prevent explosive access to the magazine. Delayed arming shutters have a rotary action which initially provide the metal barrier during firing and then brings the explosive train into line at the appropriate time. Examples are shown in Fig. 3.

Firing Devices

Firing devices ensure correct initiation of the fuze and consist of a striker, firing pin or needle designed to impinge on the detonator. A variation on this device is the inertia pellet. This is a metal weight containing a detonator which rides on to a striker or it may contain a striker which rides on to a detonator. Examples are shown in Fig. 4.

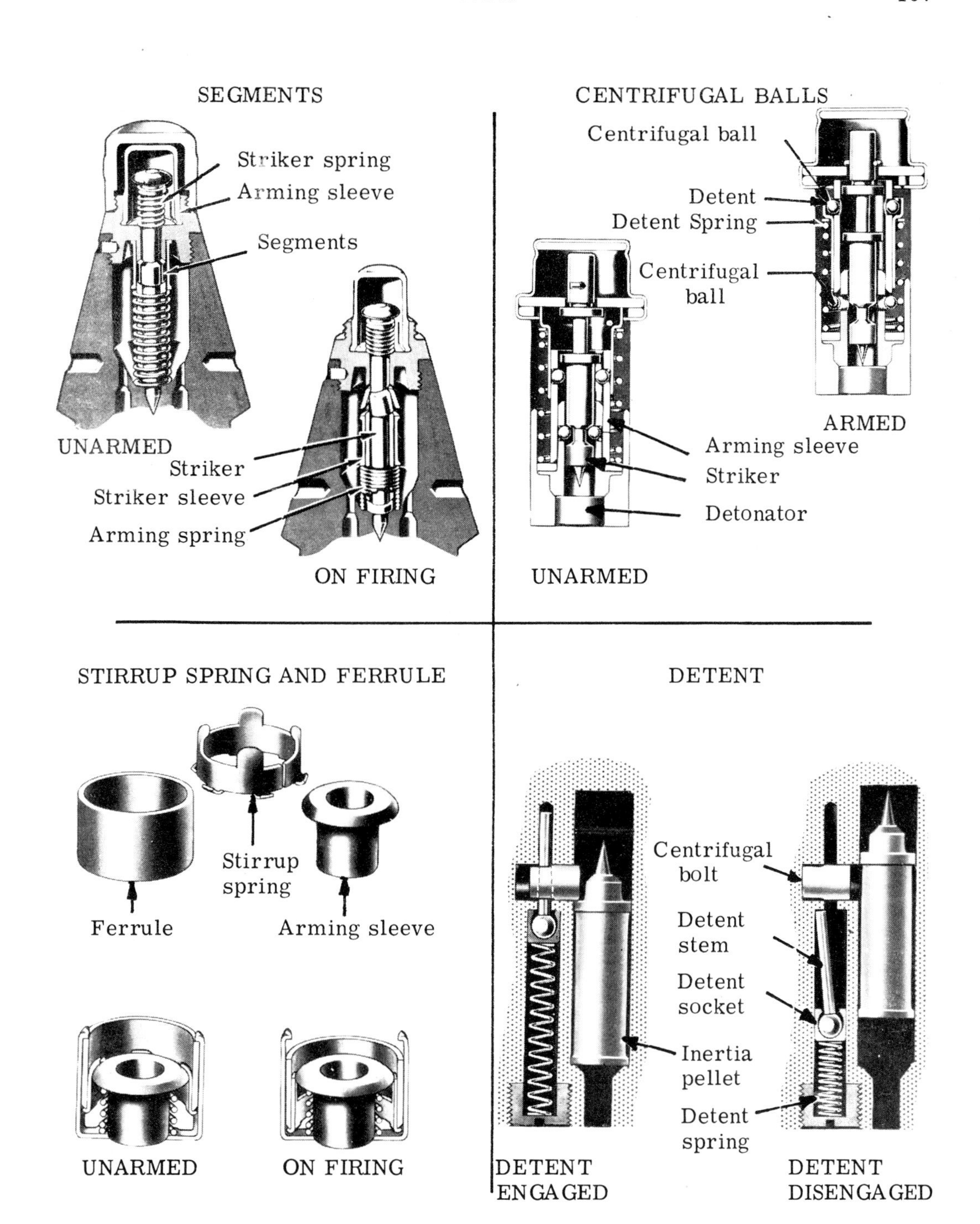

Fig. 2. Holding devices

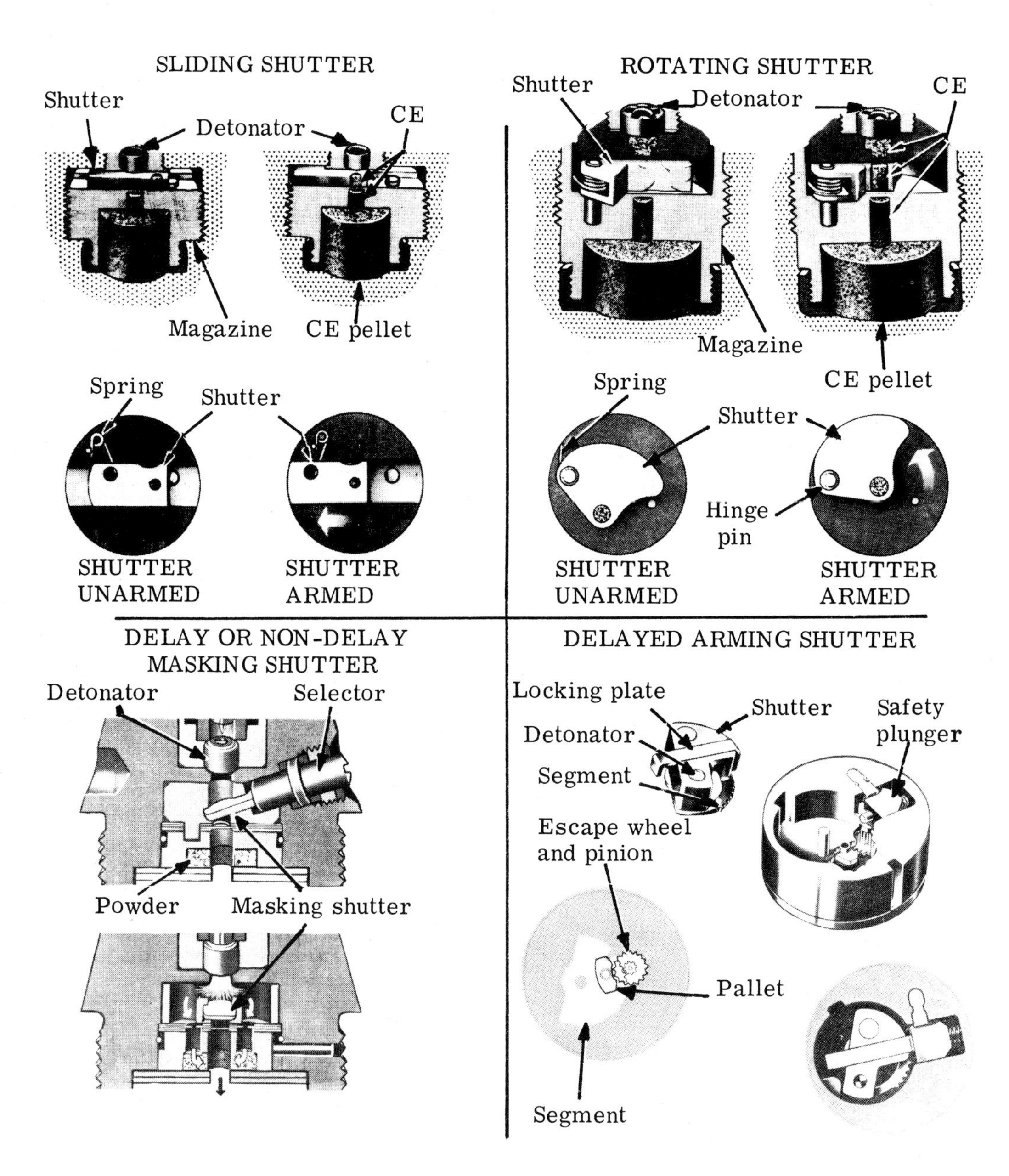

Fig. 3. Masking devices

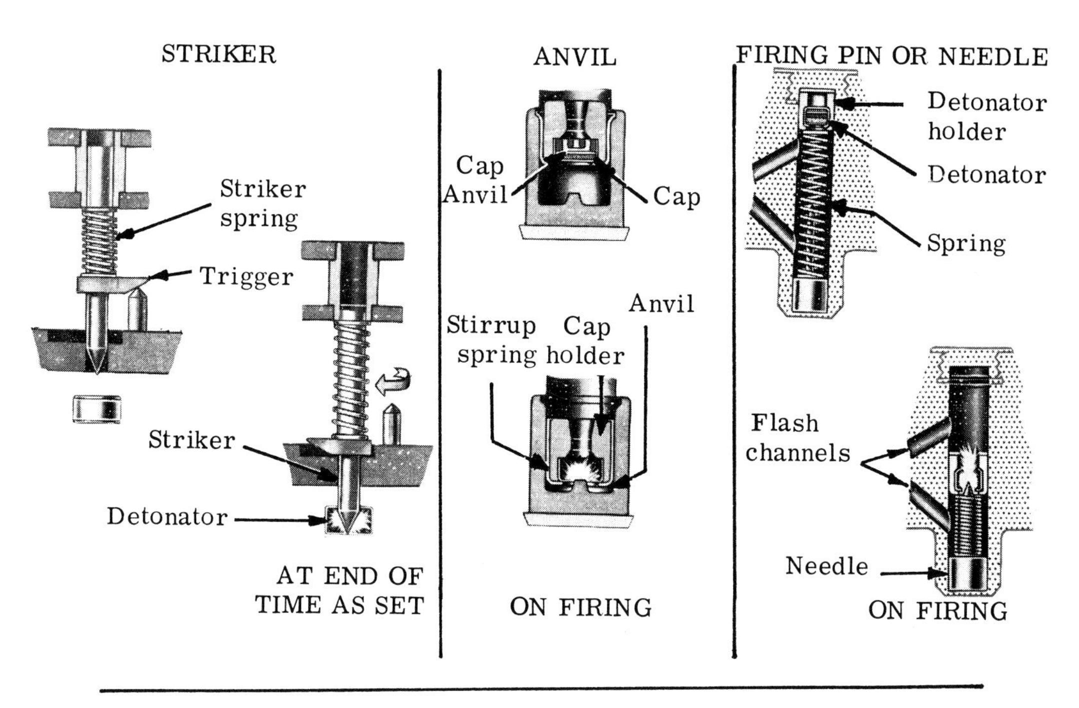

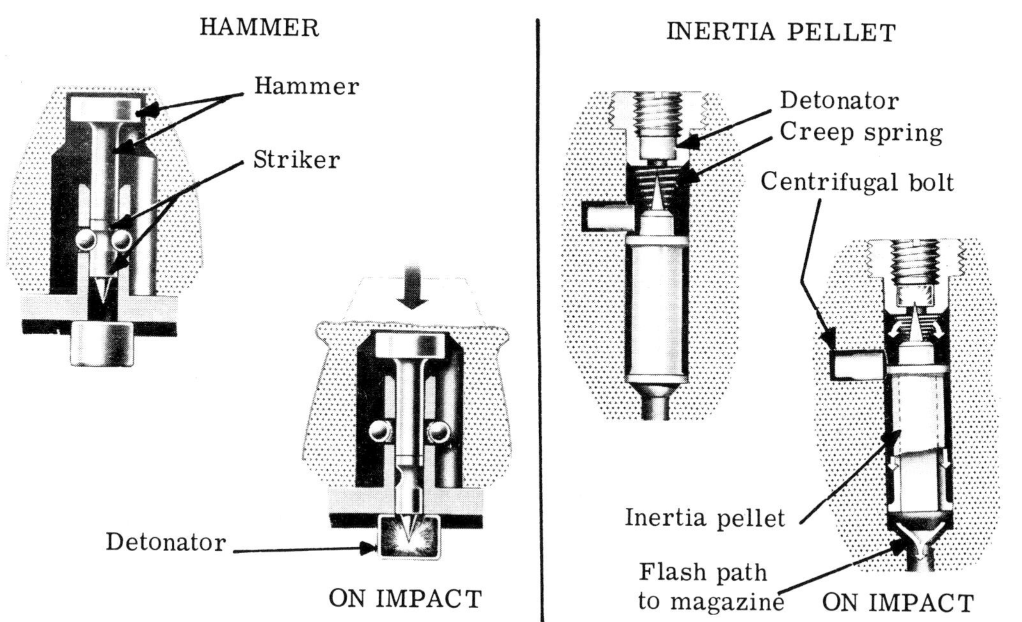

Fig. 4. Firing devices

Other Fuze Components

Apart from the mechanical devices referred to above there are various components used in fuzes to complete the explosive train. These include detonators, stemming, pellets and magazines. Detonators are usually containers filled with one or more types of explosive; stemming is found in connecting channels and ensures continuity in the train; pellets are usually made from prepressed explosive and used to reinforce flash or cause a delay; magazines may be separately attached to the fuze body or form part of it and they contain explosive which links the initiator system to the main filling of the projectile or main store. Examples of some of these items are shown in Fig. 5.

DETONATORS

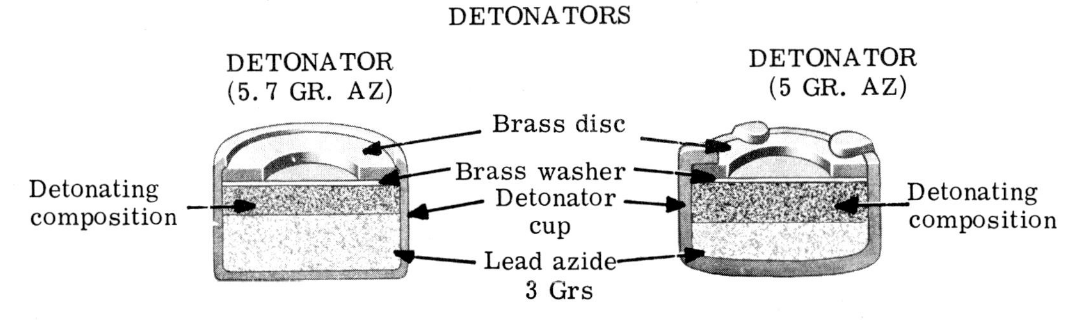

MAGAZINE AND CHANNELS

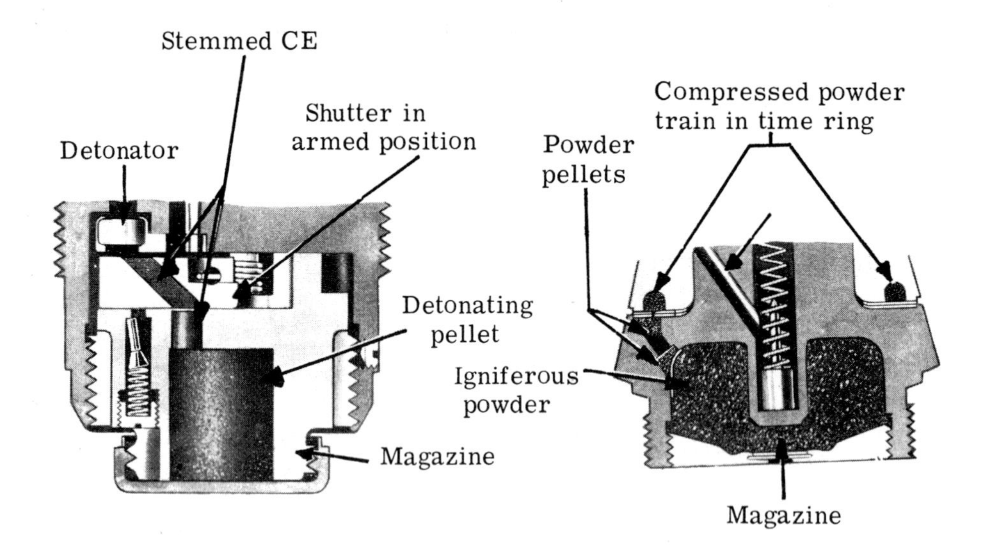

Fig. 5. Additional fuze components

TYPES OF FUZES

Time Fuzes

These are capable of operating on predetermined times and the time can be set on the fuze by hand or automatically before loading into the weapon. The process involves the movement of a ring relative to the body of the fuze and usually shows the amount of movement by graduation marks embossed around it. There are two basic types of time fuze - Combustion and Mechanical. Combustion fuzes rely on the burning of a train of compressed pyrotechnic powder which burns at a constant rate. The powder is contained in circumferential grooves in adjacent time rings, the powder burning in one ring until it can proceed through to the other. Mechanical fuzes rely for their timing on a clockwork mechanism which arms the fuze at the preselected time. Clearly mechanical time fuzes are more accurate than combustion types and are tending to replace them. Examples of both types of fuzes are shown in Fig. 6.

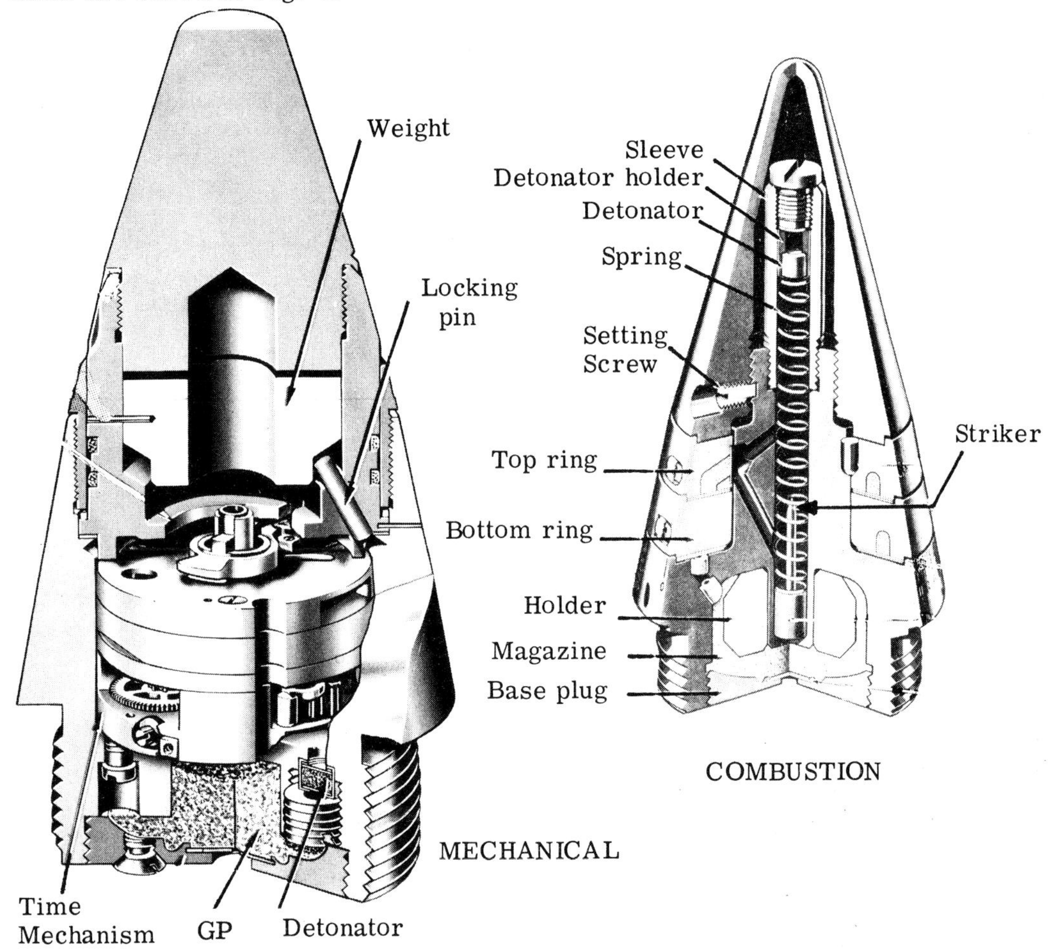

Fig. 6. Time fuzes

Percussion Fuzes

Percussion fuzes operate on impact with the target or by considerable retardation. They may be designed to function after impact to allow a degree of penetration. Most of the mechanical devices used in these fuzes have been discussed earlier and fuzes under this category can also be referred to Direct Action (DA), Graze or Delay Action Graze. Other interpretations may also be encountered such as Point Detonating (PD) and Post Impact Delay (PID) but these are not used by British designers. Direct action types are usually the fastest percussion fuzes, graze is next followed by delay types. A self-destruction device is fitted in fuzes used in the air target role so that in the event of a target miss the projectile detonates in the air and not on the ground. An example of a percussion fuze is shown in Fig. 7. Although not strictly in this category but commencing its operating cycle on impact with the target is the 'spit back' fuze used in certain HEAT projectiles dealt with in Chapter 8.

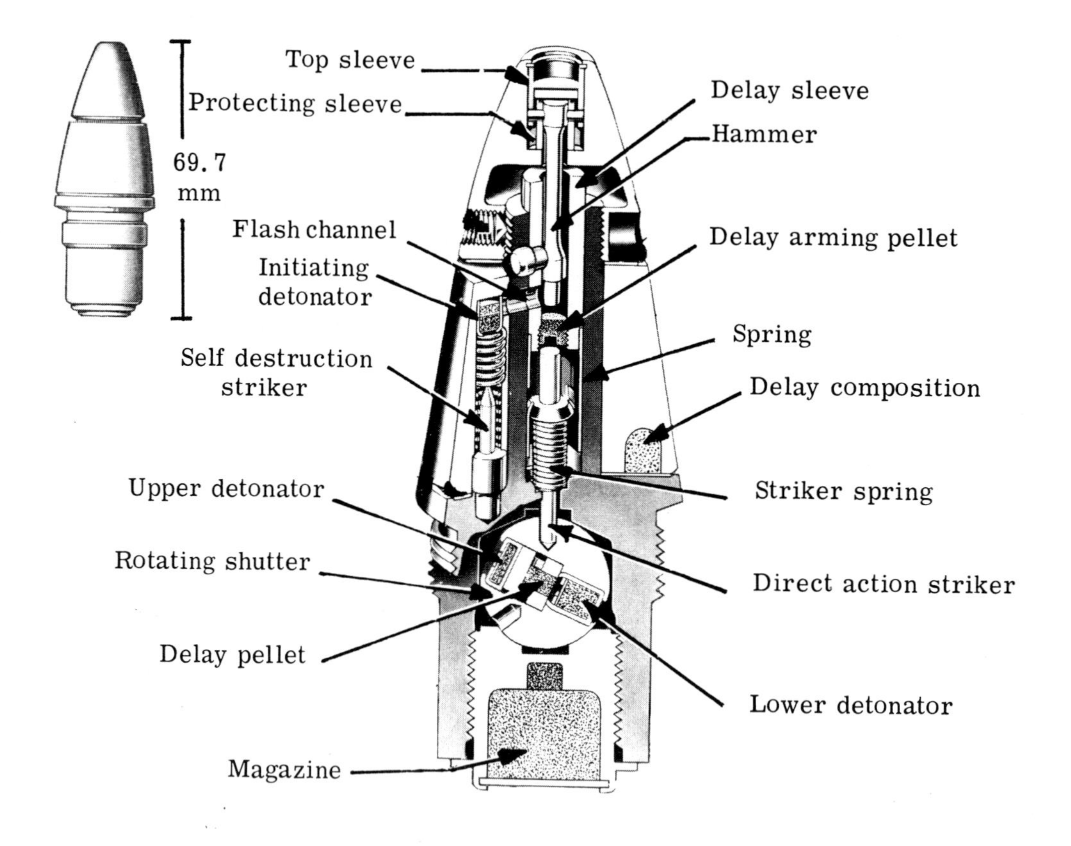

Fig. 7. Percussion fuze

Time and Combustion Fuzes

Time and combustion fuzes are time fuzes with a percussion head and mechanism incorporated.

Base Fuzes

Base fuzes are classed under the percussion type of fuze and operate on a detonator in an inertia pellet riding forward onto a fixed needle (or vice versa) when the projectile hits the target. A delay may be incorporated. They are only used now in HESH projectiles. An example is at Fig. 8.

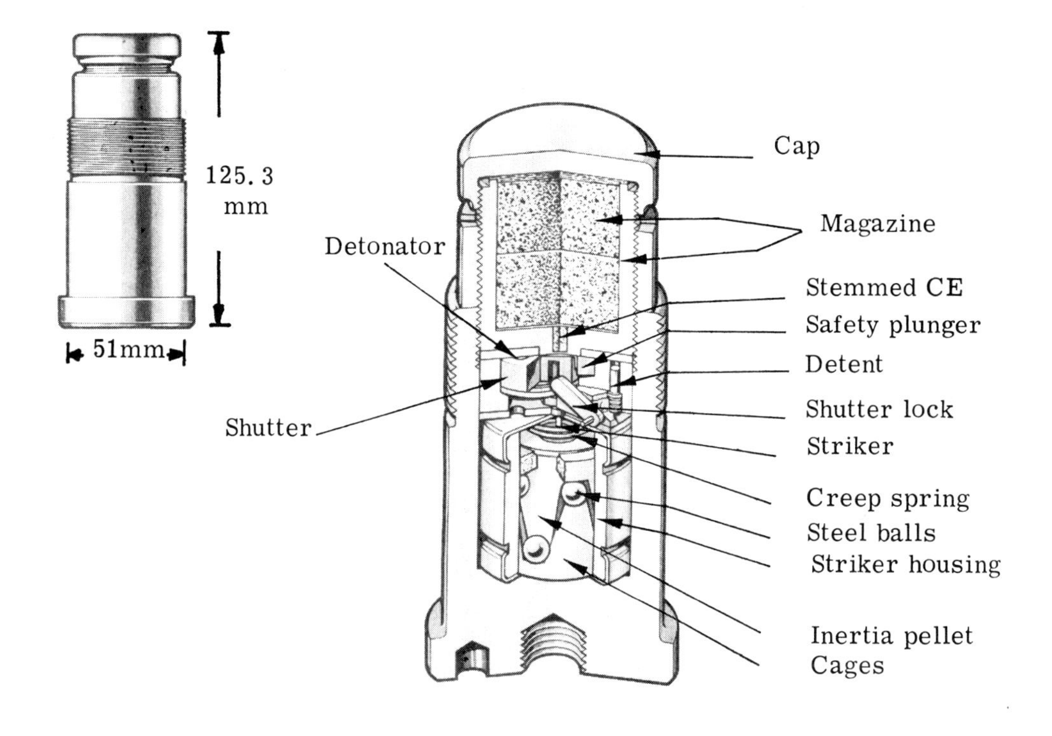

Fig. 8. Modern base fuze

Proximity Fuzes

Proximity fuzes are automatic time fuzes and are designed to operate when a transmitted signal is received back at a certain predetermined intensity. Present proximity fuzes, originally referred to as variable time (VT) are essentially a combined self powered radio transmitting and receiving unit linked to electro and mechanical devices which operate at optimum lethal distances from the target. An example is at Fig. 9.

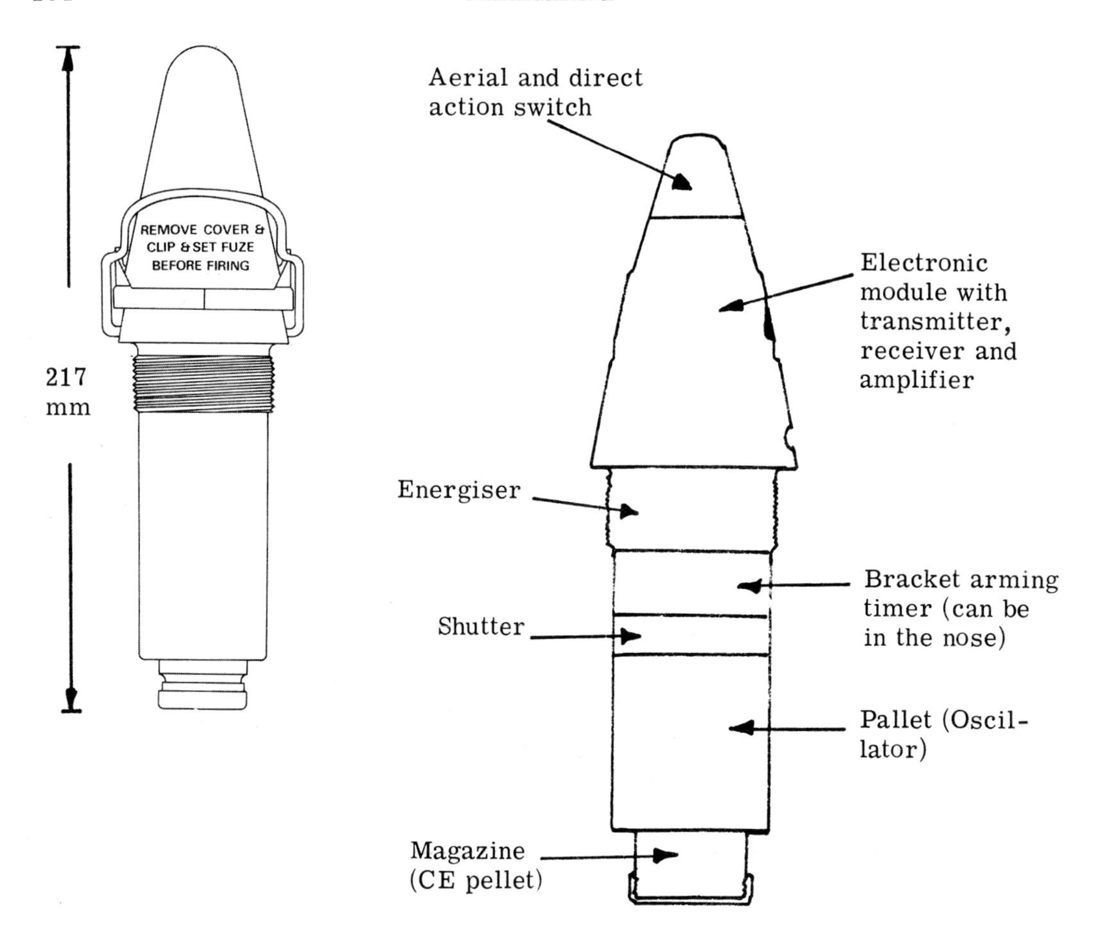

Fig. 9. Proximity fuze

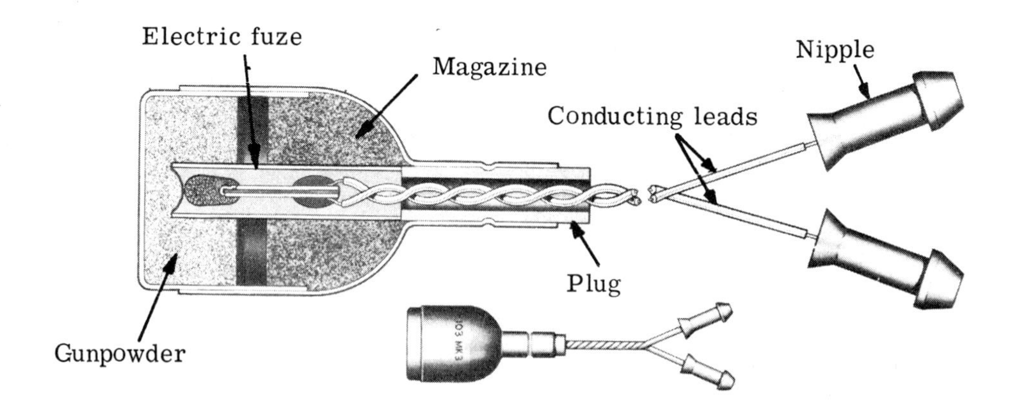

Fig. 10. Electric fuze

Electric Fuzes

Although electric fuzes do not have the mechanical and other devices fitted in most fuzes they are included for completeness. Electric fuzes are usually functioned by means of a current passing through electric leads connected to some form of battery. An example is shown at Fig. 10.

Contact Fuzes or Pressure Fuzes

Contact fuzes or pressure fuzes are normally actuated by pressure when target makes contact. They are simple in design due to the static mode in which they are used. Their main current use is with mines and an example is at Fig. 11.

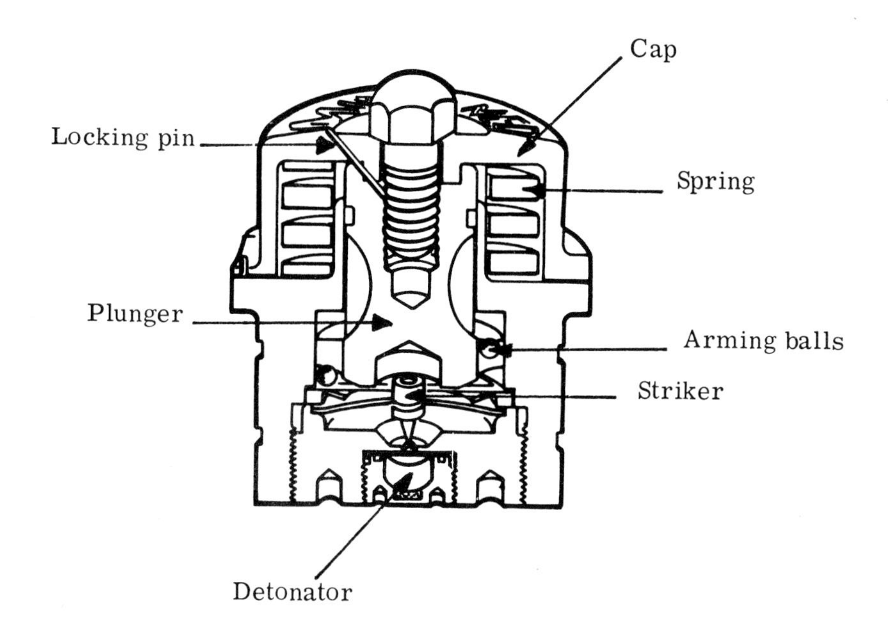

Fig. 11. Contact or pressure fuze

CONCLUSION

Several hundred fuzes have been used in the past, many of which are now obsolete or obsolescent. The tendency now is to opt for multi-role fuzes to give greater flexibility of use and to reduce the user problems in battle conditions. New materials are constantly being sought for fuze components and current design effort is directed towards reliability and reduced cost. The ability of the electronic component manufacturers to miniaturise components to adapt modern circuit assembly techniques and potting techniques have led to improved reliability and weight saving. All fuzes will undoubtedly have one or more of the safety devices referred to in this chapter and this becomes more and more important as weapons systems continue to gain improved performance.

SELF TEST QUESTIONS

QUESTION 1 What is a fuze?

Answer ....

QUESTION 2 State the basic requirements of a fuze.

Answer ....

....

QUESTION 3 How are fuzes classified?

Answer ....

QUESTION 4 Outline the forces acting on a spun nose fuze.

Answer ....

....

QUESTION 5 What are the main groups of mechanical devices used in a fuze?

Answer ....

....

QUESTION 6 What are the OB design safety principles designed to achieve?

Answer ....

....

QUESTION 7 Describe the other components likely to be used in a fuze.

Answer ....

....

....

QUESTION 8 What would you expect to find in a time percussion fuze?

Answer ....

....

....

QUESTION 9 Explain briefly the function of a proximity fuze.

Answer ....

....

QUESTION 10 What fuze would you expect to see fitted to an anti-tank mine?

Answer ....

....

ANSWERS ON PAGE 257

12

Mortar Ammunition

INTRODUCTION

Mortars were initially used as seige weapons to lob missiles over defensive works at short range targets which could not be reached by guns. When rifled weapons were introduced the mortar was somewhat relegated as it was and is a smooth bore weapon. It was reintroduced in World War II as an important weapon in trench warfare and it has changed little in basic concept since then. A mortar is a simple weapon, usually smooth bore, projecting a bomb into the air on a predicted trajectory. It is no longer however quite as cheap and simple as implied and although handling is relatively simple the weapon and ammunition designs have become quite sophisticated. The basic difference between a mortar and a gun is the mode of absorbing recoil. In a mortar the main recoil force is transmitted directly to the ground through the baseplate whereas in a gun it is absorbed by the recoil system.

WEAPON CHARACTERISTICS

In order that ammunition for mortars is fully understood it is necessary to appreciate the main characteristics of the weapon which influence the design of mortar bombs. Although mortars are covered in detail in Volume II it is worthwhile noting certain basic features in this chapter.

Method of Loading

All mortars used by the UK forces are muzzle loaded. Certain countries do have breech loading types but they are not dealt with here. Muzzle loading simplifies the design and enables the barrel and breech piece to be easily connected. Its effect on ammunition is the reverse in that the bomb must be quickly and easily loaded by having sufficient clearance between bomb and barrel yet have adequate obturation on firing.

Type of Barrel

With a muzzle loading system it is usual to have a smooth bore because a rifled barrel is complicated to load and delays the loading process. A smooth barrel can be thinner than a rifled type thus making it lighter. All UK mortars are smooth bore.

Operating Pressure

A mortar is required to be a lightweight, portable weapon. It follows that it can only operate under relatively low pressures compared to a gun. Very low charge weights are used to provide the short range requirement and relatively small charges provide the required propulsion for high angle fire up to the maximum close support ranges.

Firing Stresses

Due to the low pressure generated in the barrel on firing, the stresses on weapon and ammunition are much less than in a gun. This leads to a thinner wall in the bomb, possibly improved fragmentation and lighter weight. Lower stresses, combined with a non spun bomb, create problems in fuze design particularly from the safety aspect.

High Angle Fire

Its high operation gives the mortar bomb a long time of flight allowing several bombs to be in the air at any one time and provides improved fragmentation due to the angle of descent of the bomb.

Stability

Smooth bore weapons necessitate fin stabilised ammunition as opposed to spin stability and this poses certain problems during firing because the fin must not be distorted or degraded in the process.

AMMUNITION DESIGN

Mortar ammunition can be considered a complete round as it is loaded in one piece. The bomb, when loaded, contains the warhead, propellant charge and initiator and, when fired, leaves the chamber completely clear for another bomb to be loaded. Because there is nothing to extract or eject, the rate of fire can be high and this is an important characteristic of mortars. A typical mortar bomb is shown in Fig. 1. The types of bomb are high explosive, smoke and illuminating. Other bombs have been produced from time to time such as those attached to wire with a grapnel for trip wire functioning, but these are not now used. Practice and drill bombs are also produced for training and instruction. Current

British mortar systems are 51 mm and 81 mm only and these have replaced the earlier 2 inch and 3 inch weapons.

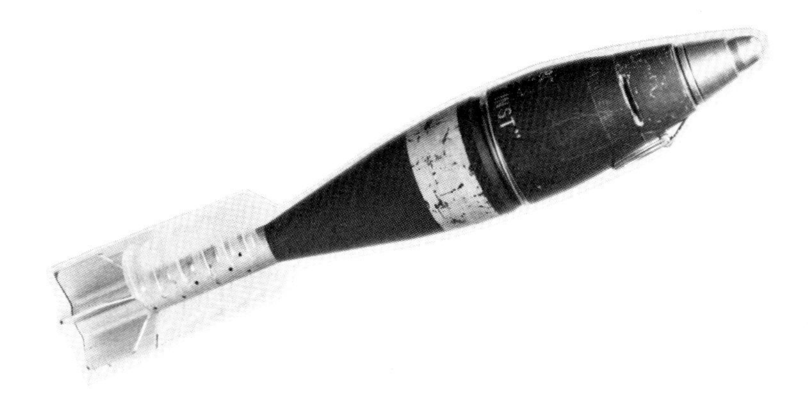

Fig. 1. Typical mortar bomb

AMMUNITION DESIGN IMPROVEMENTS

The main factors which have tended to make mortars less consistent than guns are poor obturation, large (relatively) weight and size tolerances on the bomb, fin deformation, charge arrangements and general cheapness.

Poor Obturation

Poor obturation is caused by the inability of the bomb to provide an efficient gas seal when travelling up the bore. A bomb with a close fitting tolerance would not easily or quickly slide down the bore when loaded. A guide band was machined on the bomb at maximum diameter and, although this improved obturation it did not completely solve the problem, particularly where the 'small' bomb was matched to the 'large' and worn barrel. It was also necessary to be able to load speedily in all temperature ranges and this made the problem more difficult. Modern bombs have expanding plastic obturating rings which lie flush to the bomb body in a groove when loaded but expand outwards when under pressure from the propellant gases thus effecting a good seal. The obturator ring normally drops off on leaving the barrel.

Tolerances

Weight, distribution of weight and dimensional tolerances were the natural result of what was then considered to be a cheap design. Bombs were expected to be fired in large numbers and many million cast iron bombs were made and fired during World War II. Machining was kept to a minimum and tolerances were wider to reduce production costs. Concentricity tolerances were also large for the same reason and added to the general inaccuracy of the weapon system. Modern bombs are made to tighter tolerances and have a close tolerance on the tail bomb attachment to ensure both components are in line - this has led to a more expensive bomb.

Fin Deformation

Fin deformation was caused by the method of attachment of fins to the tail tube or boom. The method used was to weld the fins to the tube in pairs and they were not easily attached in a lined up position. Fins could also be damaged or bent prior to loading. This contributed to inaccuracy which was further aggravated where fins became detached. Figure 2 shows an early tail design.

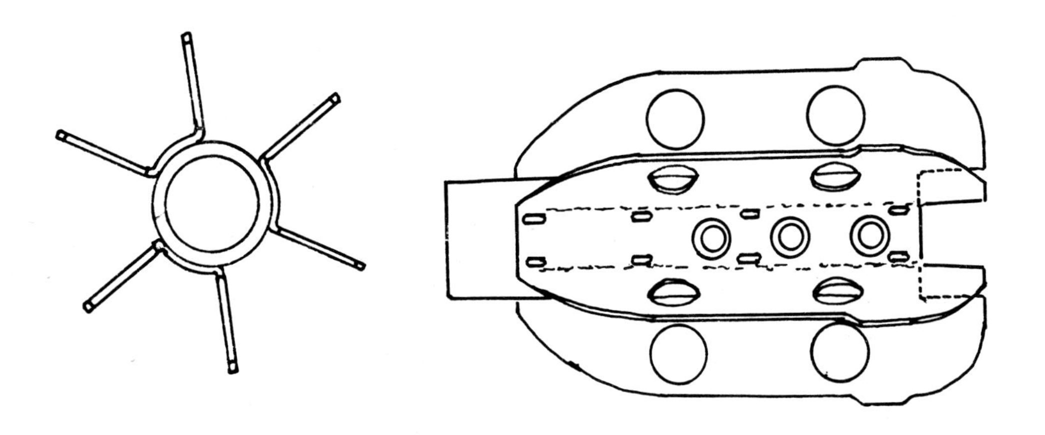

Fig. 2. Early tail design

Modern bombs have one piece tails in which boom and fins are made from one piece of metal (see Fig. 3).

Charge Arrangements

The primary charge fitted into the boom was paper wrapped, rather like a shot gun cartridge and could swell up if damp or wet, despite being protected by a varnish coating, thus affecting its performance. Secondary or augmenting charges (not fitted to 51 mm bombs) were packed in cylindrical celluloid containers and attached to the tail by a spiral spring. The celluloid container tended to crack mainly due to the chaffing of the spring. The resultant propellant loss could not be readily detected. This led to short ranging. Figure 4 shows an example of this charge system.

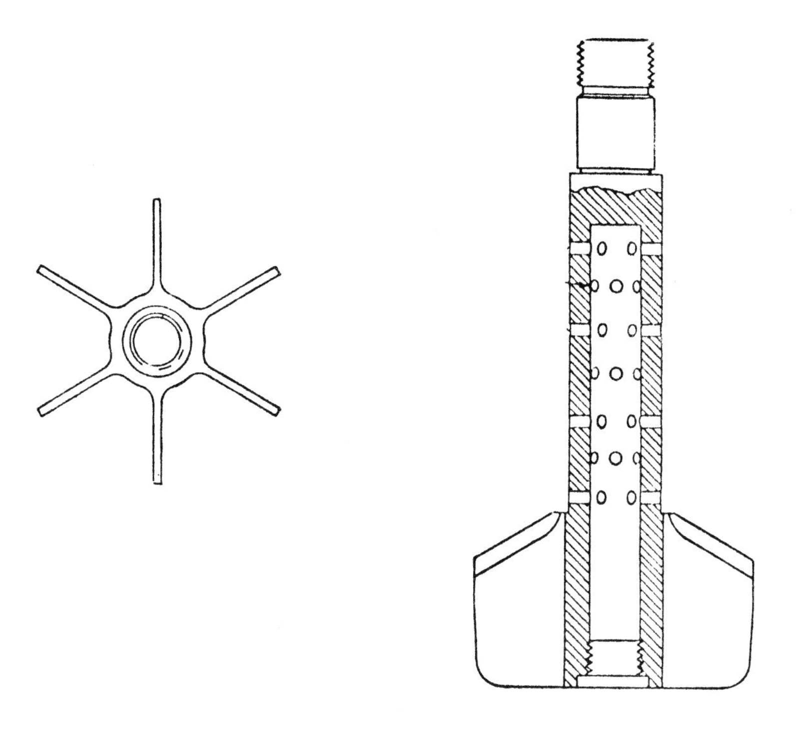

Fig. 3. Modern tail design

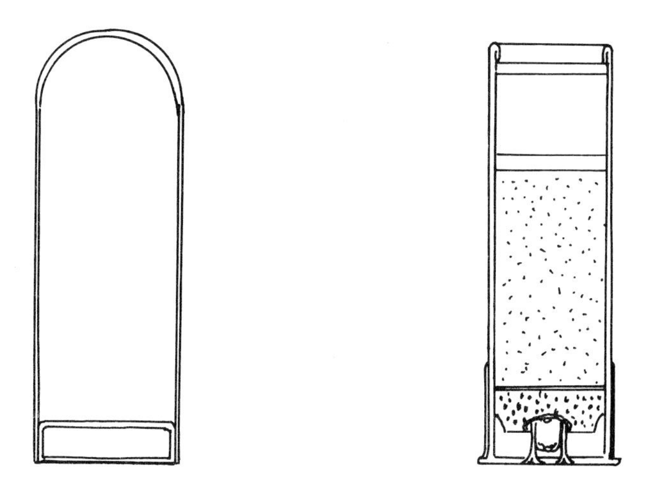

Fig. 4. Early charge system

A - M

Modern cartridge designs consist of an aluminium primary cartridge body which screws into the tail plus several horseshoe shaped secondary charges made of celluloid which clip over the boom. Figure 5 shows an example of a modern charge system.

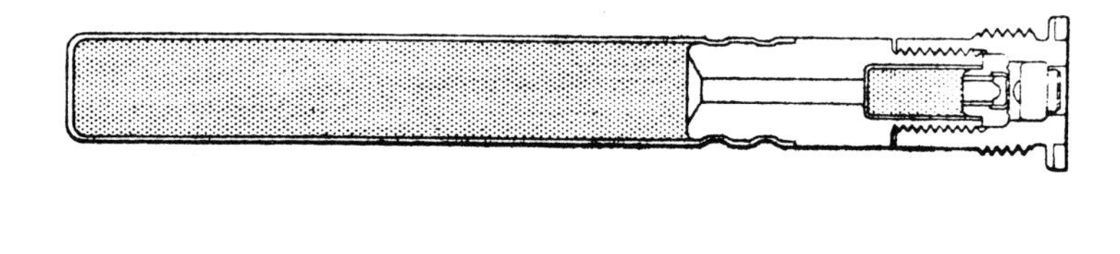

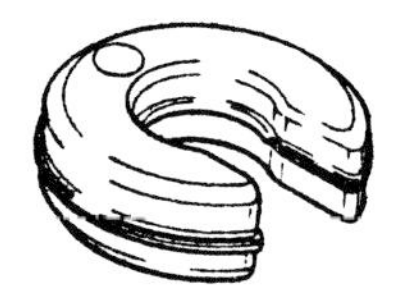

Fig. 5. Modern charge system

Fragmentation

In addition to the above improvements designed into modern mortar bombs, which have improved accuracy and consistency, there is also an improvement in terminal effects. Since the early cast iron bombs produced large fragments, only some were optimum size anti-personnel fragments whilst others were inconsequential size fragments with much dust and, therefore, they were not very effective. With the introduction of modern high explosives and improved bomb body materials the fragments now produced are more effective and consistent in their terminal effects. Small mortar bombs are constructed using a thin skin covering a notched wire coil to give pre-determined fragment sizes and shapes.

Fuzing

Fuzes are required to comply with similar design safety principles as other fuzes but, due to the smooth bore in which they are fired, centrifugal force is not available therefore "long travel" detents are often used to utilise set back forces. Safety pins are normally used to prevent premature arming and these are removed prior to loading. Some fuzes have shapes which are not aerodynamic but this is less significant in subsonic bombs than it is in supersonic types.

COMPARISON BETWEEN OLD AND NEW BOMBS

Figure 6 shows a general comparison between old and new bombs. The shape is more aerodynamically suitable and the amount of high explosive has been increased. Although the cost has also increased by a factor of three the bomb has been transformed from the 'cheap and nasty' area type effect of World War II vintage to a sophisticated, accurate, cost-effective bomb not very different in terminal effects and dispersion from that of a close support field artillery gun.

TYPES OF BOMB

There are several types of bomb used in the UK Service and these include High Explosive, Smoke, Illuminating, Practice and Drill. There are differences in design and brief details are given below. Although the descriptions refer to the 81 mm, other bombs of different designs may be encountered.

High Explosive Bomb

The high explosive bomb is made from malleable cast iron, accurately machined. The nose and tail ends are recessed and screw threaded to accommodate an adapter and tail unit respectively. A groove is machined round the body to form a seating for the obturating ring. The obturating ring is plastic and the adapters are steel or aluminium alloy. Tail units are extruded from light alloy. The bomb is filled usually with RDX/TNT and the explosive train is similar to an artillery projectile. The propelling charge consists of a primary cartridge and six augmenting cartridges which can be removed and replaced as necessary, depending on the range required.

Smoke Bomb

The smoke bomb is similar to the high explosive bomb but it is stepped at the tail end to assist identification. It is filled with white phosphorus which provides the smoke on initiation by the fuze.

Illuminating Bomb

The illuminating bomb consists of a two piece body containing an illuminating flare and parachute. At a preset time along the trajectory the fuze actuates the system by separating the body ejecting the flare and parachute which is then deployed to illuminate the area required.

Practice and Drill Bombs

Practice and drill bombs are used for training and instructional purposes.

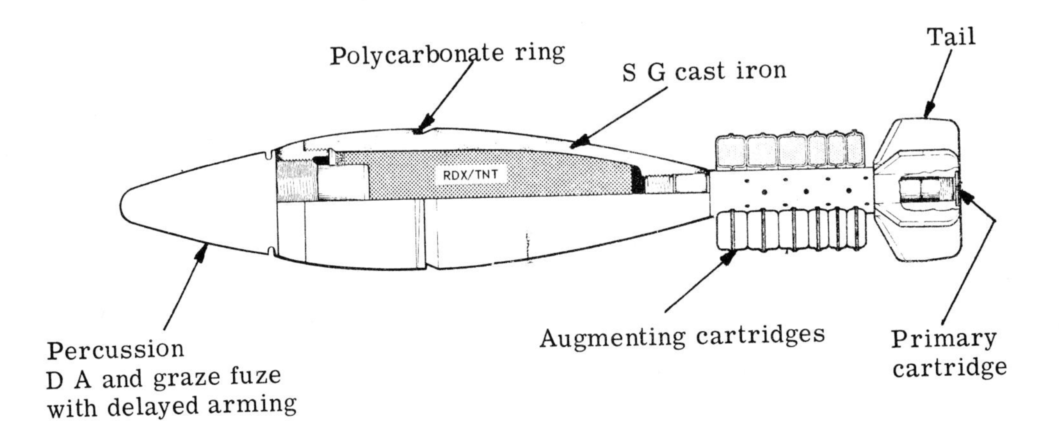

Modern 81 mm mortar bomb

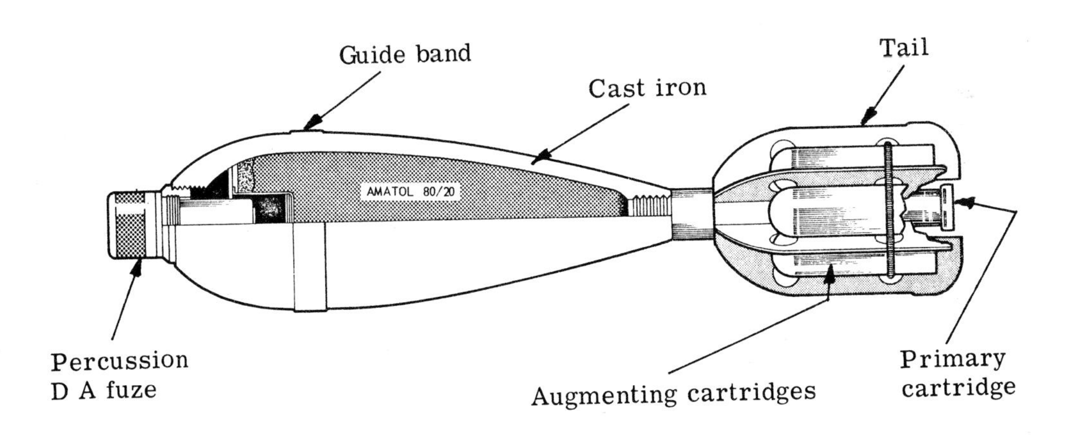

Obsolete 3 in mortar bomb

Fig. 6. Comparison of 3 in and 81 mm mortar bombs

CONCLUSION

Mortars seem likely to remain as the prime means of local fire support for the infantry. The introduction of more precise and sophisticated weapons and ammunition confirm this. Current mortar weapon systems offer greater accuracy and efficiency over relatively long ranges compared with their predecessors.

SELF TEST QUESTIONS

QUESTION 1 What is the basic difference between a gun and a mortar?

Answer ....

....

QUESTION 2 How are mortar bombs stabilised in flight?

Answer ....

QUESTION 3 Explain the need for both primary and secondary charges in some mortar bombs.

Answer ....

....

QUESTION 4 How is obturation effected with mortar bombs?

Answer ....

QUESTION 5 List the types of mortar bomb available in British Service.

Answer ....

....

QUESTION 6 How has the designer overcome the problem of fin deformation?

Answer ....

QUESTION 7 What safety devices are there in a mortar bomb fuze?

Answer ....

QUESTION 8 Explain how a relatively high rate of fire can be obtained with a mortar.

Answer ....

....

....

QUESTION 9 What advantages has malleable cast iron over normal cast iron in HE bombs?

Answer ....

....

....

QUESTION 10 What changes have transformed a mortar from an inaccurate area weapon to one which is nearly as accurate as a gun?

Answer ....

....

ANSWERS ON PAGE 258

13
Small Arms Ammunition

INTRODUCTION

The term includes ammunition used for machine guns and carbines, rifles and pistols normally associated with calibres below 15 mm although other classifications may be used such as calibres below 40 mm. Small arms ammunition is usually referred to as a round of ammunition (Fig. 1) but sometimes it is also referred to as a cartridge which means the same thing in this instance.

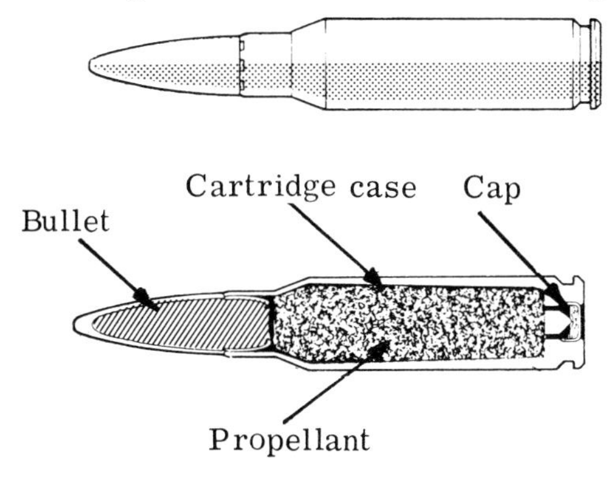

Fig. 1. Typical round

HISTORY

Small arms ammunition began circa 1320 when black powder was introduced and, for 500 years, musketeers loaded the projectile (bullet), powder and primer separately into the weapon. In the 19th century the production of mercury fulminate and a primer cap meant that a combined cartridge arrangement could be used. A major improvement was later brought about by the introduction of a metal cartridge case and, although it increased the weight of a round and led to complications in mechanisms on weapons such as extractors and ejectors, it improved many facets of small arms weaponry. Designers are now trying to turn the wheel full circle and produce a caseless round.

The bullet development is more involved. Fat slow bullets were the first in line and had calibres about 10-15 mm. 1880 saw jacketing of the lead bullet to reduce lead fouling of the bore of the weapon. The jacket consisted of a tougher metal which enclosed the lead core (Fig. 2). The jacket is now referred to as the envelope. Calibres became smaller .303 in (8 mm) in UK and Germany. Jackets were made thicker to prevent the core from being blown out of its jacket. Because calibres decreased it was thought that "stopping power" would also be affected and several attempts were made to remedy this by designing the bullet to splay out or flatten on soft targets. The superintendent of Dum Dum Arsenal, India, cut off the tip of the standard bullet to improve its stopping power but the Hague Convention in 1900 ruled it out except for "uncivilised" enemies only! There is a constant move to reduce the calibre and thus volume and weight of ammunition and many countries are now opting for 5.56 mm which weighs less than half that of the .303 round.

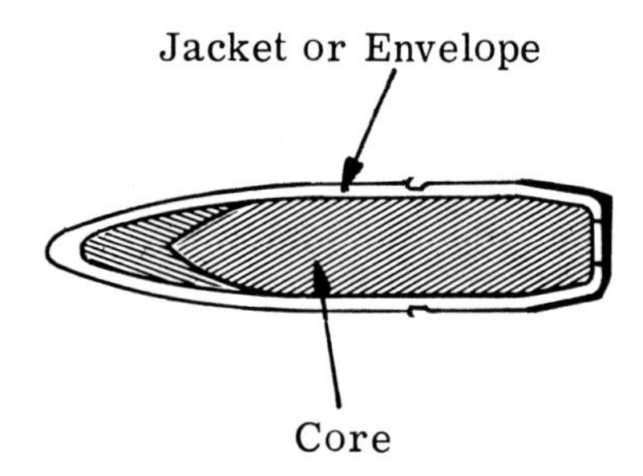

Fig. 2. Jacketed bullet

DESIGN OF SMALL ARMS AMMUNITION

Factors Affecting the Round

High velocity is required for accuracy compatible with the highest possible design pressure. Velocity needs to be maintained to maximum range, the round must not be too large nor the recoil forces too great. An increase in velocity usually entails a disproportionate increase in the weight of the propelling charge with the inevitable increase in cartridge size.

Weight and bulk should be moderate for obvious reasons, thus allowing the soldier to carry a reasonable number of rounds. Ability to stand rough usage and exposure under varying conditions is also a design requirement and this is not an easy problem to solve in view of the general size of small arms ammunition and rate of fire normally acceptable.

Safety and reliability are the other main design considerations and the true test here is when very high rates of sustained fire are needed in bad climatic and environmental conditions.

Factors Affecting the Bullet

The object is to obtain increased range and accuracy with maximum incapacitation of the target whilst minimising wear on the weapon.

Increased range depends on improving the external shape of the bullet in order to reduce resistance to the air; increasing steadiness in flight; reducing yaw and increasing the propelling charge.

Increased accuracy depends on uniformity of manufacture, so that variations in weight, dimensions and symmetry are kept to a minimum; reduced muzzle yaw and stability of the bullet throughout its flight are also essential.

CHARACTERISTICS OF BULLETS

There are two types of bullets, solid and filled. Solid bullets are used for anti-personnel and armour piercing roles whilst filled bullets comprise mainly tracer or incendiary compositions. Generally bullets have no driving bands but spin is imparted by the engraving of the envelope. The standard bullet, known as ball, presumably named after the first spherical shot, essentially consists of a metal envelope enclosing a lead alloy core. Envelope materials are usually made of cupro-nickel or gilding metal as these alloys are ductile and produce little fouling of the weapon. The core is lead antimony which gives the bullet penetration and weight and provides some "set-up" during firing thus assisting the envelope to engrave. (Set-up is defined more precisely in the glossary.)

Steadiness in flight is improved by bringing the centre of gravity towards the base. This can be achieved by using a tip of lighter material in front of the lead alloy core. Various materials have been used such as compressed paper, aluminium, compressed fibre, plastics.

To minimise air resistance, the nose of the bullet should have a high calibre radius head (ie radius of ogive should be large compared with diameter of bullet). Up to 8 gives best results. Streamlining of the base is needed where supersonic velocities are involved but for rifles, where the velocity is high up to 600 metres, it is more difficult to obtain satisfactory set-up which may cause loss of accuracy. Therefore ball for rifle use generally has a cylindrical base whilst for machine guns it is streamlined.

The armour piercing bullet is similar to the ball except that it has a toughened steel or tungsten based core. Cannelures or grooves may also be found on both types of round formed around the body near the base for attachment to the cartridge case.

Filled bullets are those where tracer or incendiary composition is used: the tracer composition is ignited when the round is fired and the incendiary when it impacts on the target. There may also be a combination of the two and these are used for observing and spotting. Examples of various bullets are shown at Fig. 3.

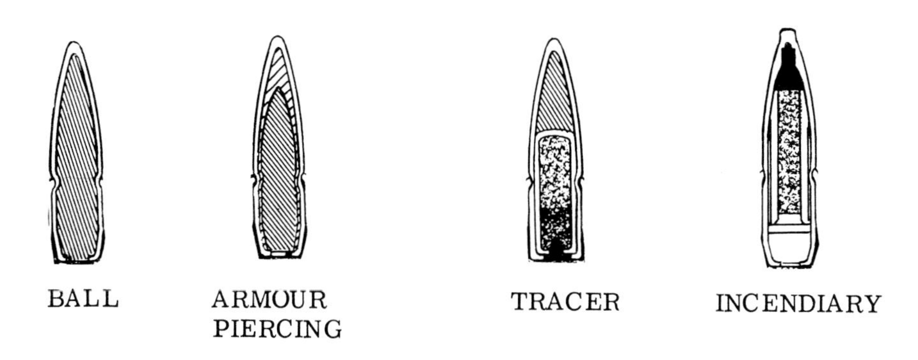

Fig. 3. Types of bullet

CARTRIDGE CASES

The cartridge case plays an important part in the small arms system and its requirements are to hold the bullet with a constant known value, provide obturation, contain and protect the propellant and initiation system. It also centres the bullet in the bore at the correct distance, provides the means of extraction and prevents erosion of the chamber, bolt face and firing pin of the weapon.

Cartridge cases are generally made from brass, although cases made from cupro-nickel, gilded metal, steel and plastic may be encountered.

A large propellant charge is required in modern high performance small arms, therefore the case diameter is usually larger than that of the bullet so that the charge can be accommodated in a reasonable length case. It is necked at the forward end to accept the bullet to which it is attached by various methods as shown in Fig. 4.

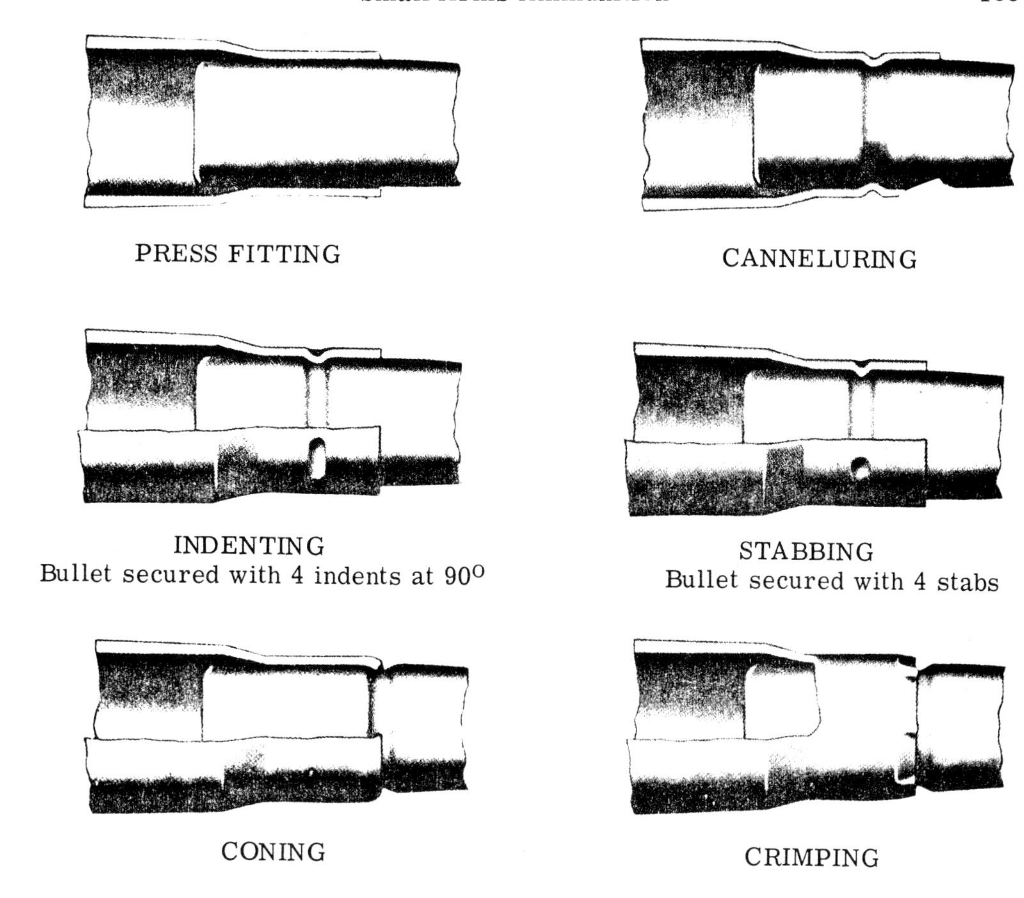

Fig. 4. Methods of securing bullet to case

A slight tapering down of the cartridge case aids extraction and thus assists a rapid firing rate where appropriate. Ammunition used in short range weapons such as pistols is loaded with smaller charges because the velocity and pressure required is lower, the sides of such cartridge cases are therefore almost parallel. Figure 5 shows a high velocity and low velocity round.

The feed and extraction mechanism of the weapon and the ignition system to be housed influences the design of the cartridge case base. Various types of base profile have been produced and are shown in Fig. 6.

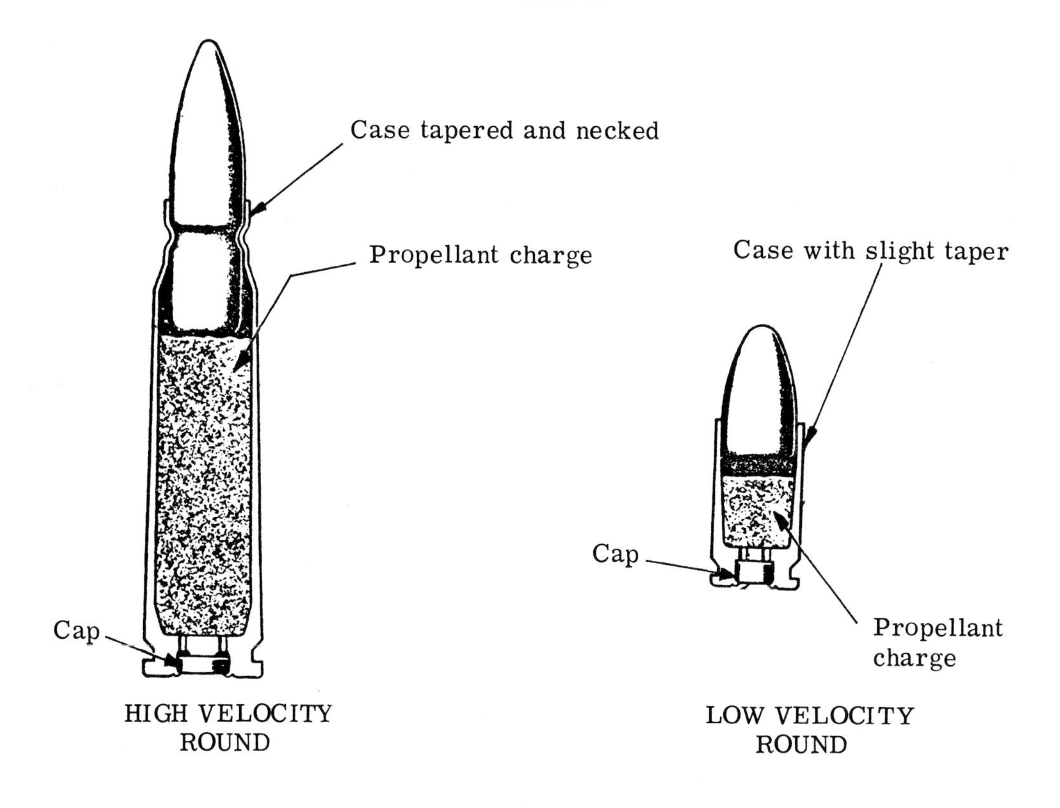

Fig. 5. HV and LV rounds

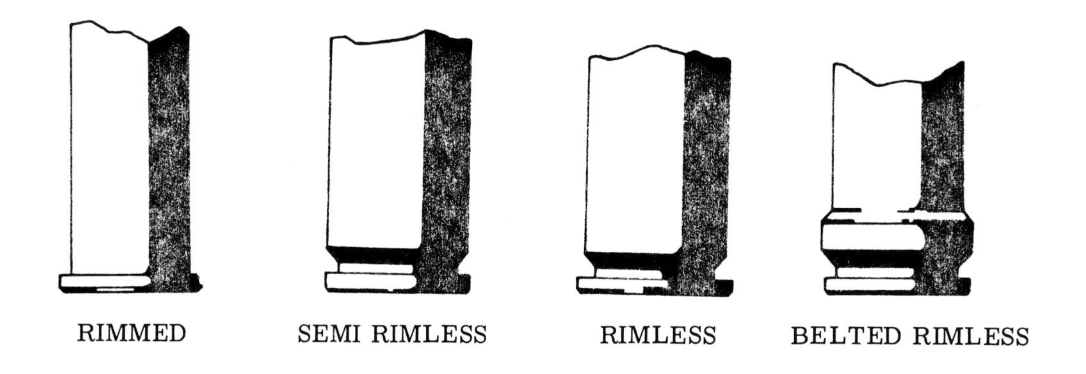

Fig. 6. Base profiles

The rimmed round is losing popularity because it is not so efficient in automatic weapons as rimless because the rim protrudes and catches on the next round when incorrectly loaded in the magazine. The rimless round is most commonly used, whilst the semi rimless and belted rimless types are seldom encountered in the British service.

The method of initiation used in small arms generally relies on a striker in the bolt of the weapon hitting on a small cap filled with sensitive explosive composition and fitted to the cartridge case base. This action squeezes or traps the composition between the cap and an anvil thus causing initiation. This produces a flash and hot particles which pass into the main cartridge case thus igniting the main propelling charge. Three types of ignition system are used; the integral anvil, separate anvil and rim fire. See Fig. 7.

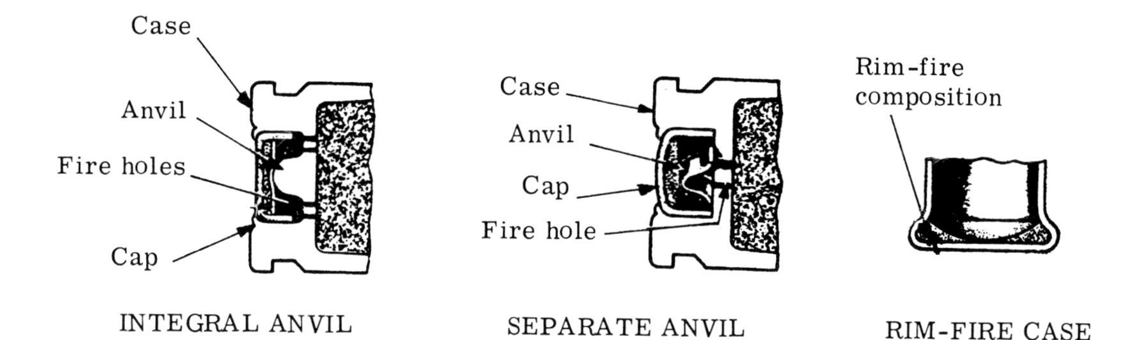

Fig. 7. Ignition systems

In the integral type the anvil is found as part of the cartridge whilst in the separate type the anvil is assembled with the cap. The rim fire system is used mainly in .22 ammunition and dispenses with the cap and anvil by the filling of initiating compositions on the inside section of the rim.

MANUFACTURE OF CARTRIDGE CASES

Each case is deep drawn from a disc, stamped from a sheet of brass. The drawing is carried out in several stages with one or more intermediate annealing treatments ending with final annealing to produce either a hardness gradient decreasing from neck to base for high velocity and uniform hardness for low velocity rounds. The soft neck is needed for speedy expansion to provide a gas seal whilst the hardened base is required to effect extraction and contain rearward pressures.

The fitting of the cap to the case is very important because it must not blow out when the round is fired. One of the following methods may be employed; press fit used for low pressure rounds such as blanks and rifle grenade cartridges; ringing, the most common method, by pressing a lip of the cartridge case over the cap edge; burring which is a special type of ringing used in very high pressure rounds such as proof rounds; punch stabbing by three or four stabs around the periphery of the base over the edge of the cap.

PROPELLANT FOR SMALL ARMS AMMUNITION

Although propellants are mentioned elsewhere in this volume, it is convenient to highlight the particular requirements and some detail of propellant used in small arms ammunition. Due to the relatively short time cycle for firing and the high firing rate usually associated with small arms, propellant must be fast burning and thus fine grained. Sliced or chopped tubular propellant is used but this can cause barrel erosion due to high flame temperature. Nitro cellulose powder such as neonites produce less wear. Single base propellants with nitro cellulose of varied nitrogen content 12.6 to 13.4% are often used. More energetic powders may eventually be introduced but they are likely to be more erosive giving shorter barrel life. Other types of propellant used include a porous grain powder in the form of discs, ballistite, mortar powders, Nobels Rifle Nemite and Nobels Nitric Neonite.

INITIATING COMPOSITIONS

Initiating compositions are sensitive compositions similar to those used in other initiators such as primer caps. Mercury fulminate was used and later potassium chlorate was added to increase flash and produce hot particles. Antimony sulphide was added still later to improve flash and increase temperature. Although the percussion ignition system is over a century old, no single explosive compound has been found which satisfies all the criteria for small arms initiation. Some caps will still be found filled with the composition referred to above and known as A1 composition, but modern caps use compositions such as VH2 which are comprised of lead styphnate, tetrazene, barium nitrate, calcium silicide, lead peroxide and antimony sulphide.

TYPES OF ROUNDS

Examples of various types of small arms rounds are shown at Fig. 8 with brief notes.

CASELESS SMALL ARMS AMMUNITION

The search for a caseless round and thus a much lighter round is not new. In 1891 a German patent was lodged called "Metal jacketed bullet with a propellant contained in a jacket lengthened towards the rear". Shortage and cost of copper in World War I led to further research into other materials for cartridges such as steel, plastic and aluminium. In World War II Germany started a study on caseless ammunition. A 9 mm Mauser rocket round was produced and several caseless types but none were very successful. USA first became involved taking the conventional cartridge case to the limit in design then looked for a caseless

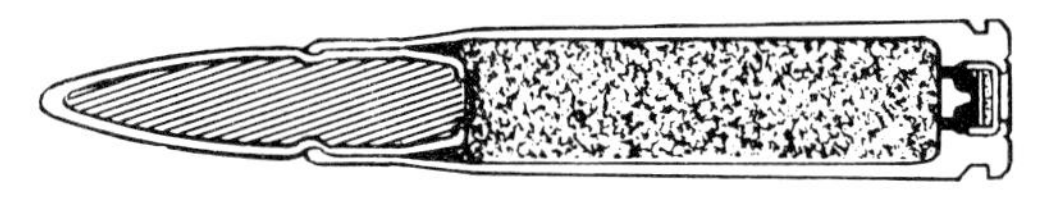

STANDARD A/PERS ROUND

BALL

TO INDICATE TRAJECTORY AND ACCURACY

TRACER

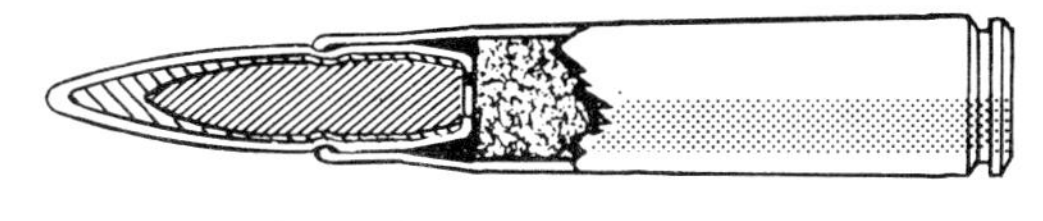

FOR PENETRATION OF THIN ARMOUR ETC

ARMOUR PIERCING

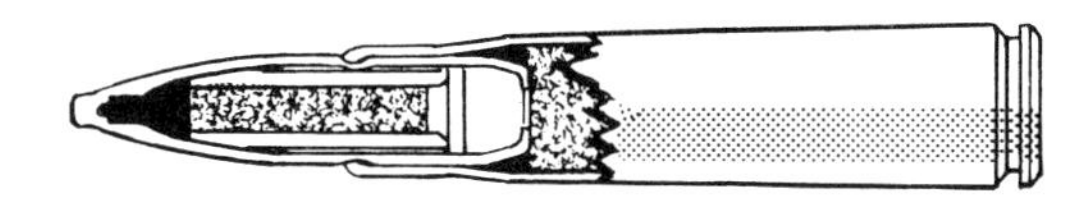

TO IGNITE AND PRODUCE FIRE

INCENDIARY

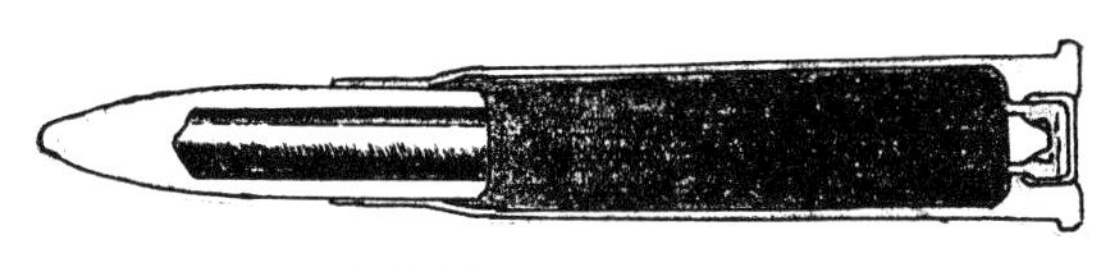

FOR FIRING IN GAS OPERATED WEAPONS FOR TRAINING

BULLETED BLANK

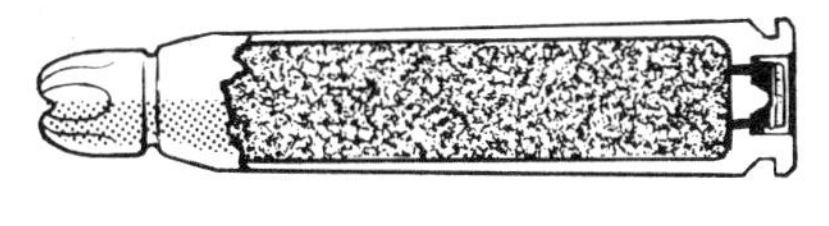

STANDARD ROUND FOR TRAINING

BLANK

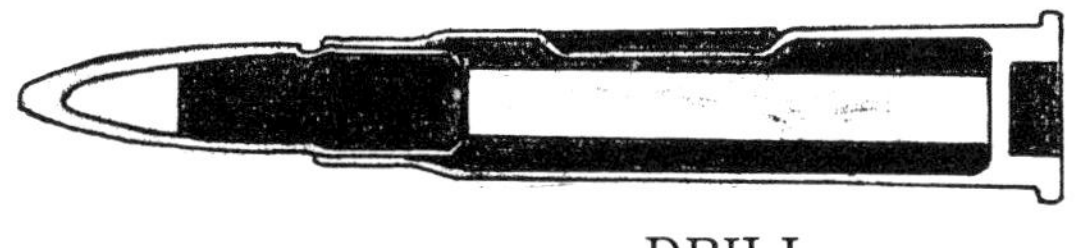

FOR TRAINING AND HANDLING

DRILL

Fig. 8. Types of round

A - N

solution. It was generally agreed that a new look was necessary and that the weapon system needed to be redesigned for caseless ammunition rather than adapting a current weapon to fire such rounds. Without the expensive brass cartridge case, ammunition should be much cheaper (30%) and lighter (50%), and work is still going on in various countries to try and find a solution. Development however has been plagued by numerous problems including: location of the initiating cap and its extraction if not of the complete combustion type; achieving consistent internal ballistics; cook-off; extraction in the event of a misfire and so on.

No doubt a caseless round will arrive in the future which will cope with all these problems and this may lighten the burden of the Infantry soldier.

SELF TEST QUESTIONS

QUESTION 1 Describe a typical round of small arms ammunition.

Answer ....

....

QUESTION 2 Why has the calibre reduced over the years?

Answer ....

....

QUESTION 3 Why is a jacket or envelope necessary?

Answer ....

QUESTION 4 What factors affect the design of a SA round?

Answer ....

QUESTION 5 Explain the uses of solid and filled bullets.

Answer ....

....

QUESTION 6 What are the requirements of a cartridge case?

Answer ....

....

....

QUESTION 7 How is a round initiated?

Answer ....

QUESTION 8 Explain the three types of initiation system.

Answer ....

....

....

QUESTION 9 What are the current problems facing the caseless round designer?

Answer ....

....

QUESTION 10 Why is there always a desire to reduce the weight of a round?

Answer ....

....

ANSWERS ON PAGE 259

14
Grenades

INTRODUCTION

Grenades are small bomb like projectiles capable of being thrown by hand or fired from a rifle or projector. They can be filled with high explosive for anti-personnel or anti-tank use or they may carry certain chemical compositions for smoke, signalling or illumination purposes. Early grenades were designed for hand throwing only and often they were provided with rod like tails and in some cases trailing tapes to assist accuracy and range. Tapes were quickly discarded as they became entangled with the equipment of the thrower on occasion. There has been and still is a wide variety of grenades for specific purposes although a few only are used in the UK Services.

CONSTRUCTION

A typical modern grenade is shown at Fig. 1 and comprises a container plus a fuzing system incorporating a safety device and a delay (these were originally referred to as ignitor assemblies, detonators, initiators and sometimes fuzes). Also present are a detonator or igniferous initiator and a payload. The payload can be high explosive for anti-personnel effect, chemical for screening and indicating, or pyrotechnic for signalling, fire raising etc. Included under chemical fillings are the anti-riot grenades used for crowd dispersing. Most of the non HE grenades can be referred to as carrier grenades in the same way as for projectiles but the variety of types is considerably less. The payload must be matched to the container for efficiency, reliability and required terminal effects and, as grenades are normally carried on the person, their construction must be such that the individual is not subjected to undue risks.

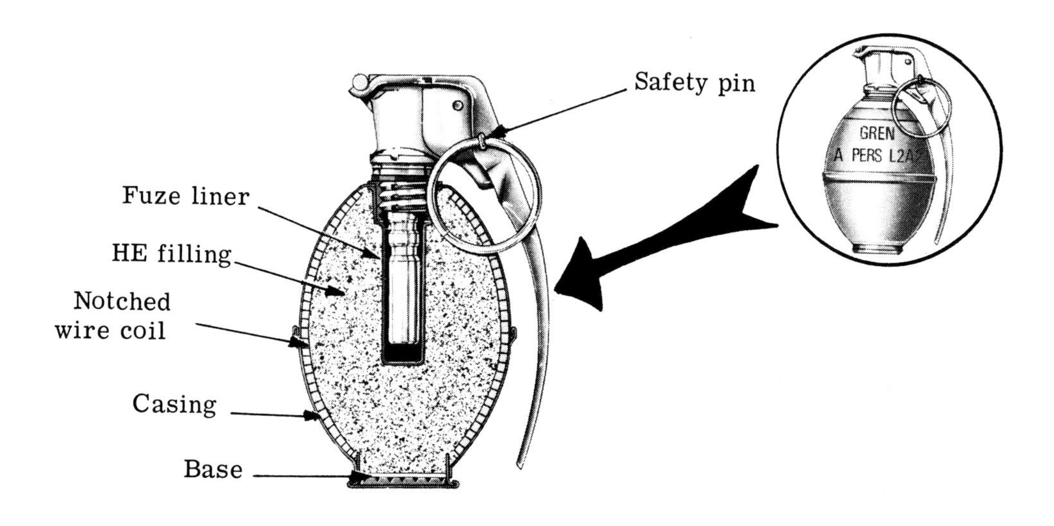

Fig. 1. Modern grenade

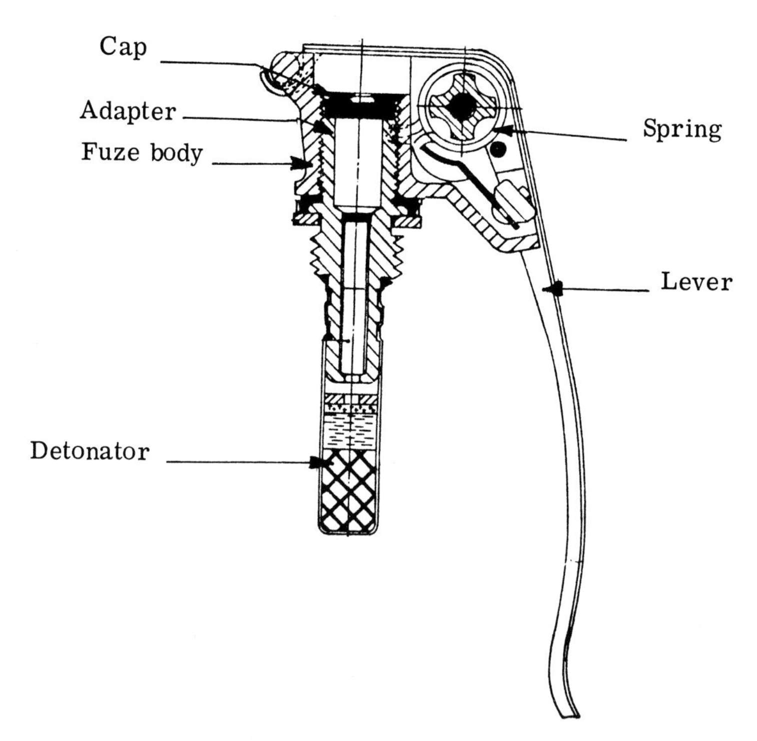

Fig. 2. Fuze/striker system

OPERATION

The fuze usually consists of a striker and safety lever plus a delaying device. A safety pin is normally incorporated and removed by the thrower or firer. The lever controls the striker and, until the lever is released, the striker cannot operate. The thrower removes the pin whilst holding the grenade and safety lever in his hand. On throwing the grenade the lever flies off allowing the striker to impinge on the initiator thus starting the delay after which time the grenade functions as designed. Figure 2 shows an example of a fuze striker system.

PROJECTION

The limited range obtainable by throwing led to the introduction of various methods of projecting grenades so that greater ranges could be achieved. Grenades were fitted with long steel rods which were inserted into rifle muzzles and fired but the recoil was too excessive and barrels became very quickly damaged. There were other shortcomings to this system which made it unpopular so that it is not now used. Special projectors were introduced for grenade delivery but this added virtually another weapon to the Infantry armoury and was not popular. The smooth bore discharger cup is another method for achieving greater range and consists of a small cup attached to the rifle where range is adjusted by elevating or depressing the rifle or by the use of gas ports which vary the rate of propellant gas escape. A rifled discharger cup may also be found which matches the rifling of the barrel: the grenade is fitted with a pre-engraved driving band. This method gives greater accuracy but there are problems such as the limit in the size of grenade used due to tendency to twist the rifle when fired. By using a spigot attachment at the forward end of a rifle a simple method for projection is available. The grenade has a hollow tail tube closed at the grenade end and this fits over the projection spigot. The tail tube becomes the barrel of the system. Finally and more recently a bullet trap type can be used where the tail tube of the grenade accepts the standard bullet when fired from the rifle to which it is attached and the energy is transmitted to the grenade thus propelling it to the required distance. Some examples of grenades for these projector systems are shown at Fig. 3.

GRENADE DESIGN

Many factors affect the design of grenades, the object of which is to convey the largest payload over a predictable range as accurately as possible. There is clearly a limit in weight and in range. Weight is a limiting factor for the thrower and for the strength of the rifle when projected. The lethal area is also influenced by the explosive and metal content of a high explosive grenade and will require the thrower to be protected in certain circumstances. Some countries adopt the policy of using "offensive" grenades relying mainly on blast effect, with a small lethal area and no protection needed for the thrower, and "defensive" grenades with larger lethal areas where the thrower must be behind cover. Ideally grenades should be capable of adaption to both conditions.

Other design requirements are safety of operation, relaibility, waterproofing, consistency and so on.

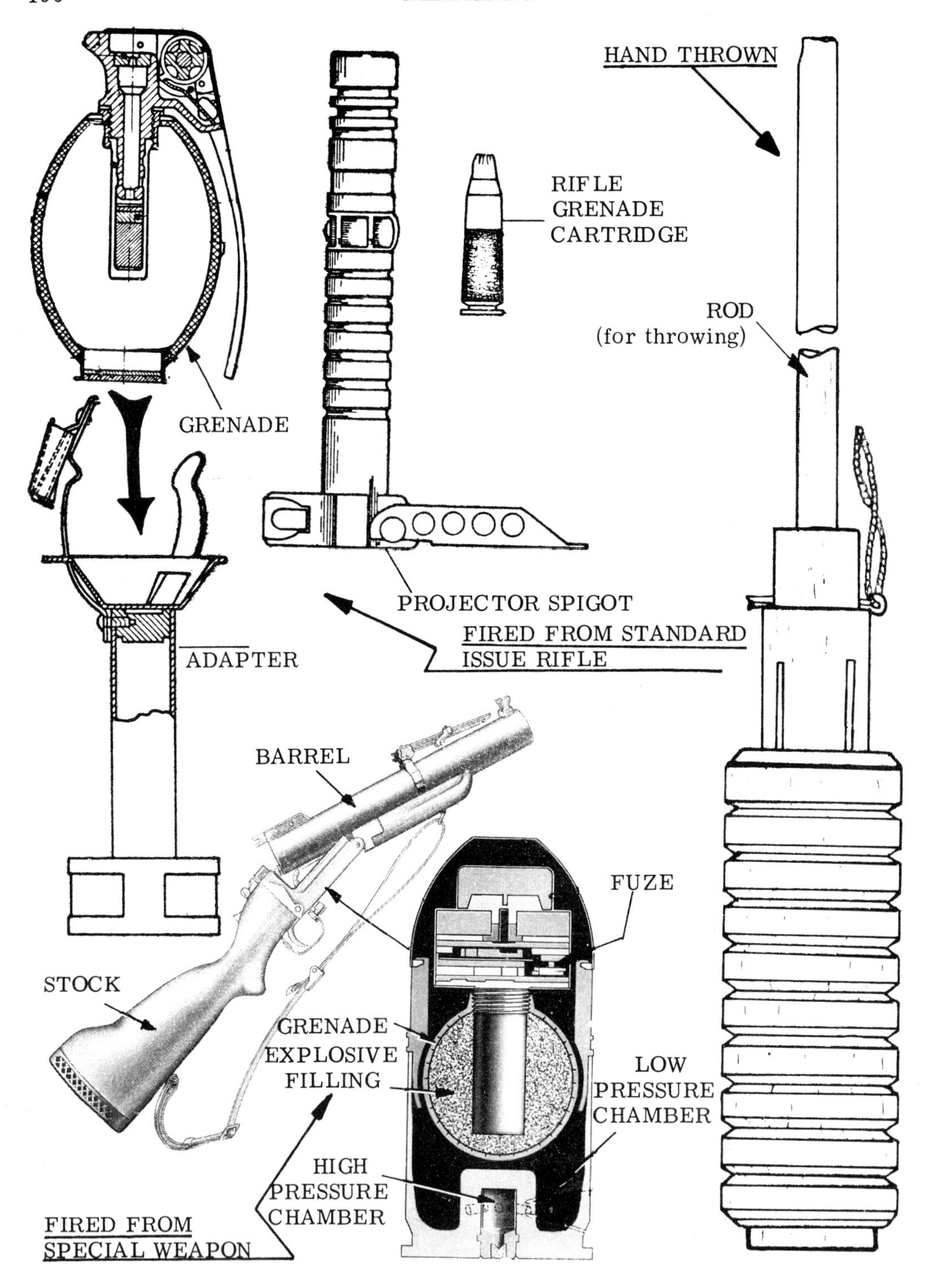

Fig. 3. Various types of projected grenades

THE FAMILY OF GRENADES

High Explosive Grenades

High explosive grenades consist of a metal body notched on the inside or a thin metal case containing a notched wire insert. The case is filled with high explosives which, when detonated by the fuzing system, propels fragments at velocities of 1000 m/s to produce casualties and damage. Another grenade not now used in the UK Service is a blast type grenade which relies on the high explosive effects only. A grenade which can be used for both situations is shown at Fig. 4 where the explosive and fragments can be separated allowing the grenade to be used in the offensive or defensive role.

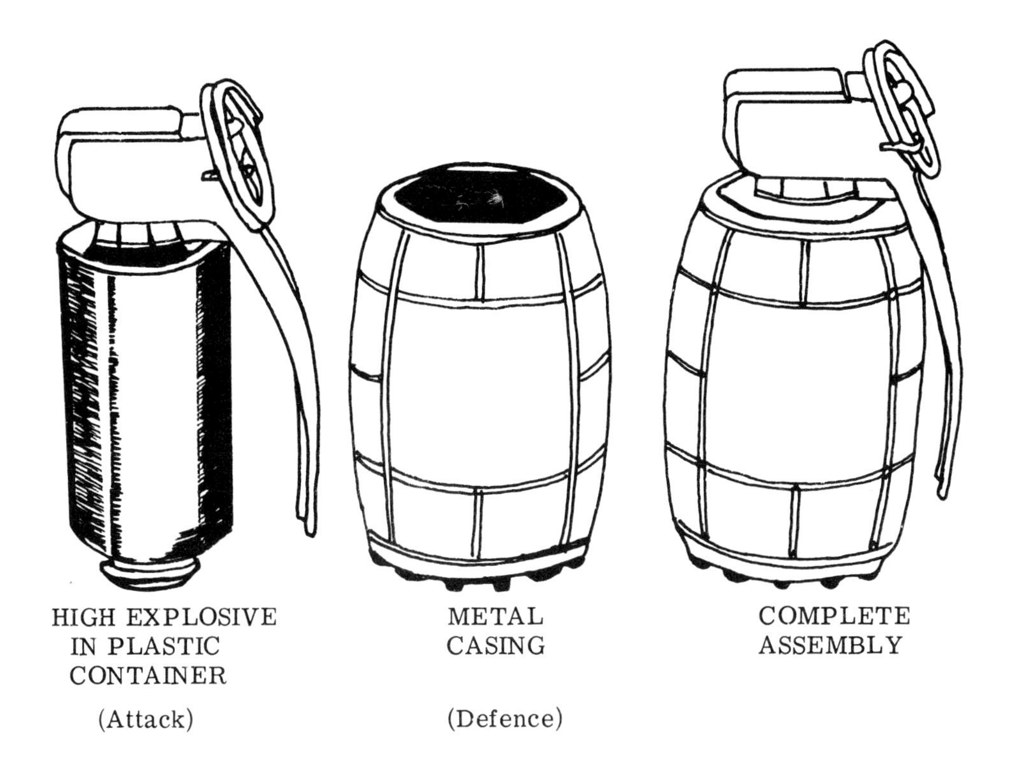

Fig. 4. Dual purpose grenade

Anti-Tank Grenades

The majority of these use a shaped charge to achieve the maximum effect against armour within strict limitations of size and weight. The grenade No. 94 (Energa) shown in Fig. 5 was the last grenade of this type to be used in the British Service. It was fired from a spigot rifle projector using a ballistite cartridge. The nose contained a direct action spit-back fuze, which fired a detonator housed in the base of the shaped charge. The maximum range was 300 metres although the hit chance was low at this range. Armour is now much thicker and sophisticated and it is doubtful whether a grenade of this size would have much success against modern armour.

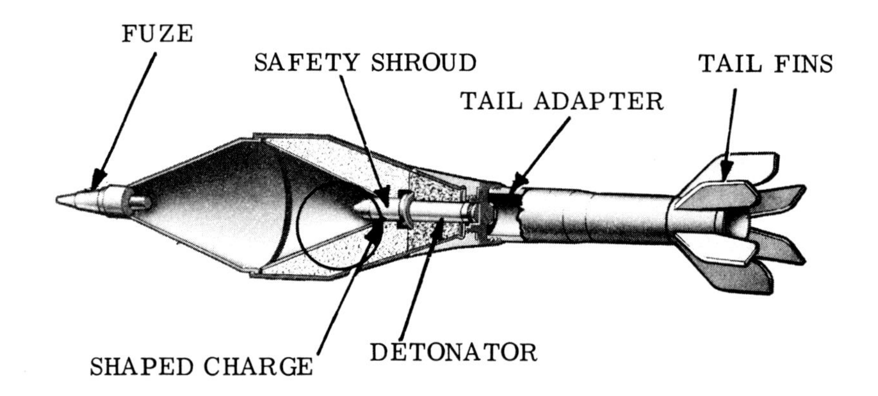

Fig. 5. Anti-tank grenade

Carrier Grenades

These contain various fillings and consist of smoke, white or coloured, and irritant gas for anti-riot purposes. Some countries use flare, incendiary and other fillings.

The modern tendency is to reduce the size of these grenades to ease the weight on the soldier who has to carry them. Fig. 6 shows this trend from the normal size No. 83 grenade to the new midi grenade.

Smoke grenades are used for screening the target when white smoke is needed whilst coloured smoke grenades are used for signalling. Colours used are blue, green, red and yellow.

Anti-riot grenades are used for dispersing rioters by producing a cloud of irritant smoke which is neither lasting or injurious.

Grenades used in Dischargers

There are two main types of discharger. One was introduced for use with fighting vehicles and mounted on the vehicle; although originally designed to enable the vehicle to provide an immediate smoke screen for short duration for its own protection its use has been extended to crowd control. The other is fired by hand from the shoulder and is a single shot, muzzle loading weapon. Fig. 7 shows an example of each type. A number of grenades are available for use in these dischargers.

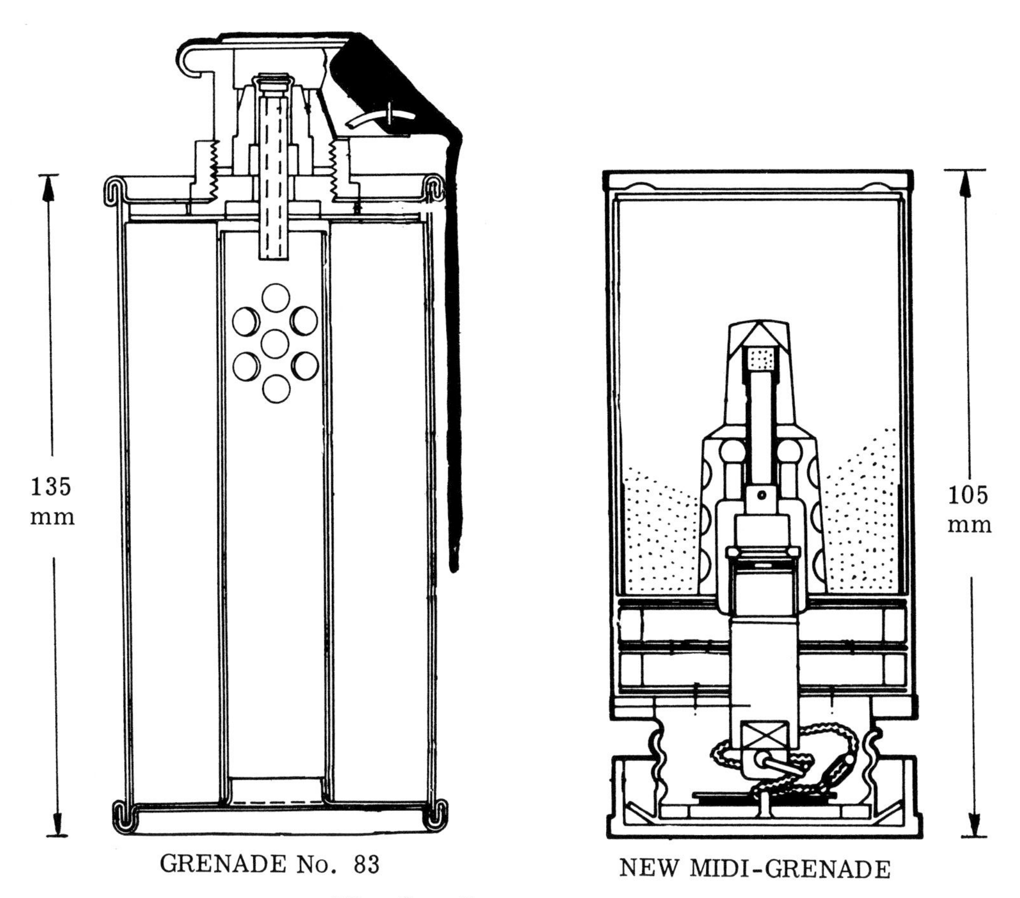

Fig. 6. Grenade 'shrinkage'

Ring Aerofoil Grenades (RAG)

In an effort to achieve greater accuracy, grenades are under development shaped rather like a streamlined (hollow) doughnut. These aerodynamic projectiles generate lift in flight giving a flatter trajectory without velocity increase and therefore longer range.

'Multi' role Grenades

Ideally a single grenade should be available for all requirements. The French have developed a grenade that can be thrown or projected. It can also be easily adapted for use in the offensive or defensive role.

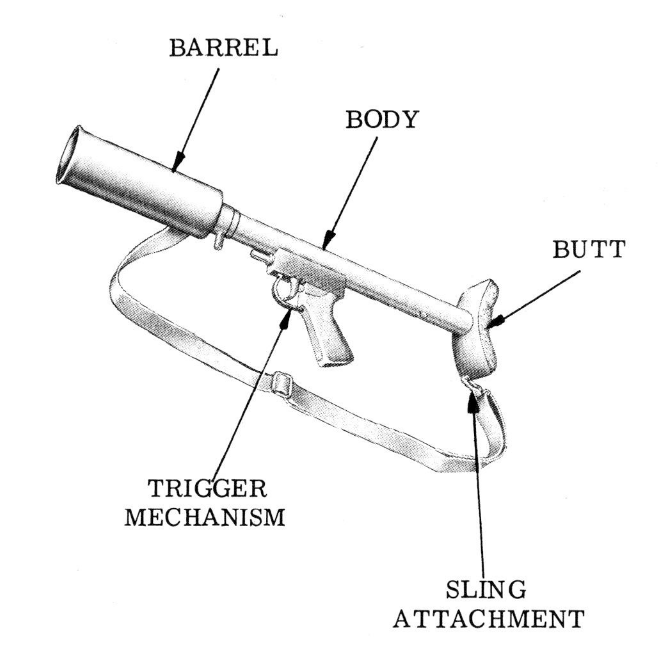

SHOULDER FIRED

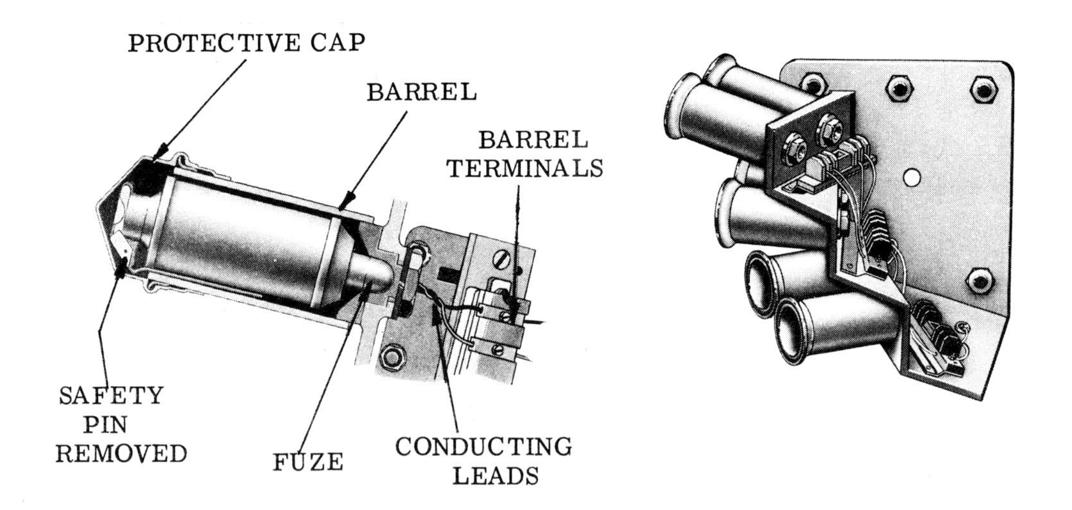

VEHICLE MOUNTED

Fig. 7. Grenade dischargers

CONCLUSION

Grenades are still an essential item of ammunition and although originally of simple design modern battle requirements and safety considerations have introduced some complications in design. Use of modern materials and techniques have led to more efficient performance and cost effectiveness and the crowd control requirement has added a new and important member to the grenade family.

SELF TEST QUESTIONS

QUESTION 1 Describe a modern anti-personnel grenade.

Answer ....

....

....

QUESTION 2 What types of payload are used in modern grenades?

Answer ....

QUESTION 3 How is a grenade made safe to handle?

Answer ....

....

QUESTION 4 Explain the various methods of propelling a grenade to the target.

Answer ....

....

....

QUESTION 5 Compare the requirements of an offensive grenade with a defensive type.

Answer ....

....

QUESTION 6 List the basic requirements for a grenade.

Answer ....

....

QUESTION 7 Why are anti-tank grenades no longer popular?

Answer ....

QUESTION 8 What is a carrier grenade? Give examples of payloads.

Answer ....

....

QUESTION 9 Explain the operation of a RAG.

Answer ....

....

QUESTION 10 What is a multi-role grenade?

Answer ....

....

ANSWERS ON PAGE 260

15

Mines

INTRODUCTION

A mine, in a munitions sense, is "a receptacle filled with explosive, placed in or on the ground for destroying enemy personnel or material, or moored beneath or floating on or near the surface of water for destroying or impeding enemy ships". This definition embraces a variety of classes of mine. Mines are usually classified by their target or the environment for which they are designed; anti-personnel mines, beach mines, anti-tank mines and so on. They are also often classified by some unusual aspect of design for example: acoustic mines, pressure mines and non-metallic mines.

This chapter, however, will only deal with the important aspects of the main types of mines used in land warfare: anti-tank and anti-personnel mines.

GENERAL CHARACTERISTICS OF MINES

Component Parts

The main component parts of a mine are the case, the fuze, the explosive train, and the main explosive system. This applies to all mines except special mines such as chemical mines.

The case is the container for the entire mine assembly. Its shape and size are of great importance, as is the material from which it is constructed. It must protect the contents in whatever environment the mine is to be used and it must also be robust enough to stand up to manual and mechanical handling.

The fuze is the mechanism by which the mine recognises that the right kind of target is above it, and it is the fuze that sends the firing impulse to the detonator at the right moment to create the maximum damage to the target. It may be simple like the foot-operated pressure fuze of the anti-personnel mine, or extremely complex, like a multi-sensor full-width attack fuze. It may also incorporate remote control, self sterilisation or self destruction elements. It usually consists of a sensor to detect the target, a safety and arming unit, and an actuator to fire the detonator.

The explosive train is the means by which the firing of the detonator is carried through the booster charge into the main explosive filling. The main explosive filling attacks the target. It does this either by blast effect, using the chemical energy of a shaped charge, or by the kinetic energy imparted by the explosive either to the fragmenting case of the mine, or to special projectiles as in the Misznay-Schardin, or other types of plate charge.

REQUIRED CHARACTERISTICS OF A MINE

Lethality

A mine must be sufficiently lethal to achieve its purpose. If it is not, it is useless and may have the most serious consequences tactically and on friendly troops' morale and confidence in their weapons, equipment and munitions. On the other hand, a mine can be over lethal: an anti-personnel mine which blows a man to bits is wasteful compared to one which merely blows part of his foot off. Both render the man unfit for further combat, but the second does so more cheaply, and imposes an additional load on the enemy's medical resources.

Sensitivity

A mine fuze must be adequately sensitive to actuate under the right target, but not so sensitive that it will explode under the wrong type of target. Typical actuation pressures for pressure operated mines are 8-50 kg for anti-personnel mines and 200-250 kg for anti-tank mines.

Safety

A mine must be safe to store, safe to transport, safe to lay and safe to arm. A mine that becomes unstable and explodes in storage, or explodes during arming or when inadvertently dropped during transportation, will never be accepted into service. Mines should desirably be safe for own troops to clear when necessary.

Ease of Laying

A mine must be easy to arm and lay. This is partly a function of fuze design, and partly of shape and size. For example the British Mark 7 anti-tank mine needs a special action to arm the mine during laying, a tedious and time consuming procedure that makes it relatively difficult to lay by machine. On the other hand the British bar mine, with single lever arming, is much easier to lay.

Reliability

A mine must function when a suitable target passes over it, and a mine which fails to actuate or whose self sterilisation or destruction elements are unreliable, will soon be discarded. Mines must also function reliably for periods of up to six months after being laid in hostile environments such as wet soil or water.

Resistance to Countermeasures

A mine should be as resistant as possible to such mine countermeasures as ploughs, flails, rollers and explosive over-pressure. It must be resistant to mine detectors and capable of being fitted with anti-handling devices, to counter enemy manual search and lifting.

Camouflage

Camouflage is dependent on the size, shape and colour of the mine, the materials from which it is manufactured and the method by which it is laid. A mine should be small enough to be easily concealed, either by camouflage or by burying, and small enough to prevent the scar, formed by mechanical burial, from being easily seen from the air or ground.

Logistic Considerations

The mine and its container should be as economical a transport load as possible. This again dictates an economical shape for the mine, and the minimum weight compatible with achieving the required effect.

Storage

British mines are designed for an in-service life of between 10 and 20 years. This means that a mine can be transported and stored in varying natures of environment for many years and then, when laid, still function with its original reliability. Unlike many other ammunition natures which are used in training, live mine stocks are rarely turned over.

Cost

Mines are used in large numbers. A modern armoured division may require up to approximately 100,000 anti-tank and approximately 100,000 anti-personnel mines at the start of operations. Mines can be made as sophisticated as the user requires, but only at a cost, and with a restricted budget more sophistication means less mines. The need to minimise costs is therefore vital, but not at the expense of safety and reliability. Mine costs are not only related to design and manufacture, but also to inspection, transportation and storage.

Delayed Arming

Delayed arming usually takes the form of a crude timing device, required to arm the mine at some time after laying. This delay might be for any period from 15 seconds to 40 minutes. It allows the layer to get well clear of the mine before it becomes armed. This kind of arming is incorporated in scatterable mines and is required in any system which remotely or aerially delivers mines, or where the fuzing system might well be actuated by the noise or other signature of the laying vehicle. The timing device can be mechanical or a combination of mechanical and electronic.

Self Sterilisation

Self sterilisation adds flexibility to the tactical use of mines, since they can be made to sterilise themselves after a known period of time after laying. It facilitates counter-attack, and also removes many of the hazards of eventual clearance. Mines are rarely in tactically correct positions after about four days of battle. Self sterilisation can either be short term, a period of hours, or long term, when the time can be measured in weeks. Mechanisms to allow short term sterilisation are usually electronic because of the increased effectiveness of modern micro-circuits. These can be robust, reliable and accurate. Longer term sterilisation can be done in many ways, but where electronic fuzes are used, controlled battery voltage decay may be employed. The battery is made to discharge over a long period and when discharged the mine fuze becomes inoperative. The sterilisation itself may take the form of rendering the fuze inert, interruption of the explosive train, or even the self-destruction of the mines. This last method ensures that the mine cannot be used against its original layer. Remotely Deliverable Mines (RDMs) are designed with a self-destruct capability.

Remote Control

Where gaps have to be left in minefields to allow routes for withdrawal, the closure of them is often a difficult and dangerous process. It is possible, however, to design a form of remote control mechanism to switch on the mines laid in the gap. Where remote control is required over long distances, electronic means are probably the only reliable way of achieving this, but it leaves the remote control open to the enemy's electronic countermeasures. It is, however, possible to send coded signals to mines that require time to counter electronically. These signals either switch the mine on or off, or with a more complicated fuze, interrogate the mine as to its state of readiness. This latter facility requires a transmitter at each mine, at considerable cost.

Remote Sterilisation

It is possible to make a tank, moving through one of its own minefields, generate a certain frequency signal which energises the neutralising circuits in the mines beneath it and in front of it for a short period of time; the mines can then automatically switch on again after the tank has passed.

SUMMARY OF CHARACTERISTICS

Any mine can be designed to have a wide variety of characteristics, but, as with any other equipment, the final arbiter is the cost. Many of the characteristics mentioned above add considerably to the cost and size of the mine, and affect the long term reliability. This is the problem that faces the designer and the user.

ANTI-TANK MINES

Introduction

On a battlefield dominated by armoured vehicles, the creation of obstacles will play a large part in any plans. Natural obstacles are not always available or severe enough to check the enemy, so these have to be augmented by man-made obstacles. Man-made obstacles can take many forms: ditches, trenches, concrete constructions, "slippery" agents and foam barriers. The most universal form of obstacle, however, and that used most frequently to augment the natural obstacle on the battlefield is the anti-tank mine. As an anti-tank weapon, the mine has the advantage of being able to attack the tank in one of its weaker spots, its underneath or "belly". In fact, anti-tank mines, in the main, concentrate on attacking two features of the tank, its tracks, and the belly armour.

The tracks are in contact with the ground and are usually 500 to 700 mm wide. The belly armour, which is relatively thin (about 20 to 30 mm) "stands off" from ground level by 400 to 500 mm. Neither of these targets is at the present time particularly resistant to mine attack. The majority of tank armour is usually concentrated to protect against conventional direct fire weapons, and therefore placed at the front and to a lesser extent the sides of the hull and turret.

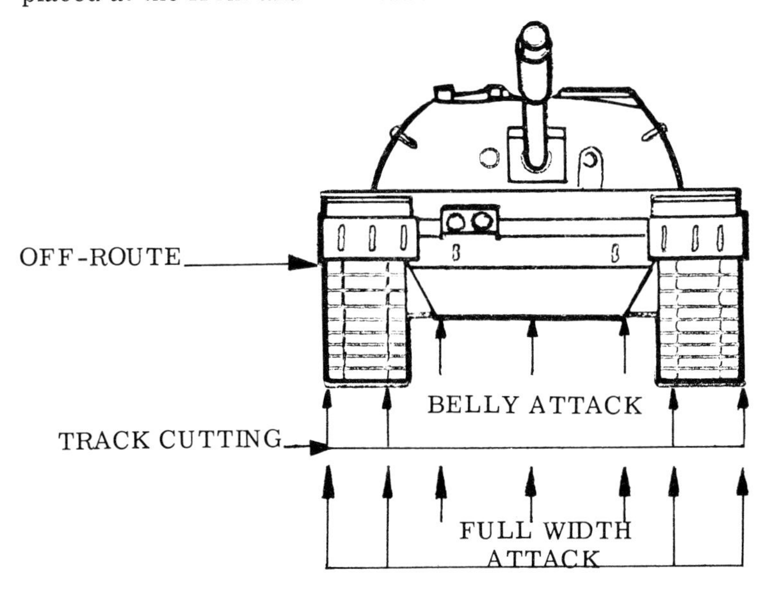

Fig. 1. The tank as a target for the mine

Types of Anti-Tank Mine

Anti-tank mines are usually classified according to the way in which they attack the tank. Those that attack the tracks or wheels of the target vehicle are categorised as "track cutting mines". Those that attack only the belly of the target vehicle are known as "belly attack mines", but are not large enough to cut the tracks. Finally, there are the types of mines that attack both the tracks and the belly. These are often belly attack mines, but should more correctly be called full-width attack mines. They are usually track cutting blast mines fuzed for belly attack, or they can be belly attack mines with sufficient explosive to produce an adequate blast effect for track cutting. Figure 2 shows typical examples of anti-tank mines.

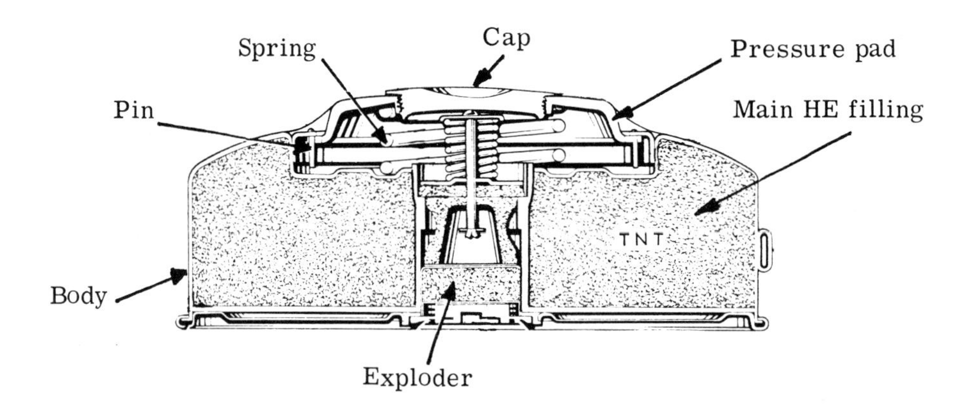

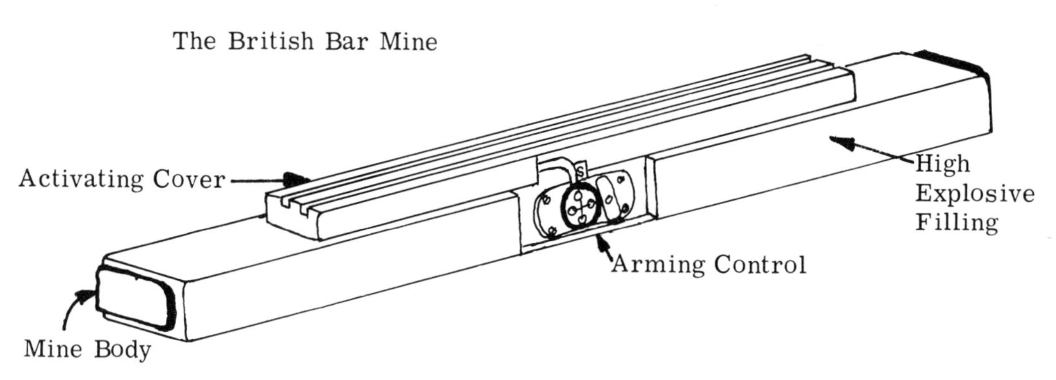

Fig. 2. Typical anti-tank mines

Off-Route Mines

Another type of anti-tank mine is the off-route mine. This type of mine is laid at the side of the likely approach route of the target, and either fires a Misznay-Schardin plate or a shaped charge through the side armour of the tank. Alternatively it can be an in service anti-tank rocket or grenade fired remotely from a fixed stand.

Explosive Content of an Anti-Tank Mine

The sizes of most anti-tank mines are dictated by either the amount of explosive they contain, or by the method of laying or delivery. The amount of explosive they contain depends on the mechanism of attack. Blast mines need at least $2\frac{1}{2}$-3 kg of explosive to disrupt tank tracks and running gear reliably. Shaped charge mines can use smaller explosive charges and this form of attack is often used where the mine must be small because of its methods of delivery. These, however, are really only suitable for the belly attack of tanks. Misznay-Schardin mines use a smaller weight of explosive than blast mines and can be smaller than them. This does reduce their capability against tank tracks and running gear.

Anti-Tank Mine Fuzes

For the track cutting mine there are three types of fuze: the single impulse fuze, the double impulse fuze, and the anti-disturbance fuze.

The simplest form of fuze is the pressure-operated single impulse fuze. This operates from the pressure exerted by the tracks or wheels of the target vehicle. The fuze or its sensor elements are normally mounted in the top of the mine and operate from a simple vertical load displacement. Such a fuze is very prone to simple countermeasures but it is cheap, and most anti-tank mines in general service in the world are of this type.

The double impulse fuze is designed to counter the simple mine roller, and operates on the second impulse on the mine. The roller takes up the first impulse, and the tank the second which detonates the mine. If a tank without rollers crosses the mine, the load of the first road wheel takes up the first impulse, and subsequent road wheels detonate the mine.

The anti-disturbance fuze is designed to counter hand lifting or mechanical disturbance of the mine by such devices as the mine plough which can be fitted to the front of tanks.

There are a number of specially designed fuzes for the belly attack mines and the full-width attack mines to ensure their maximum effect against the target. A vertical tilt rod is the simplest kind of full-width attack fuze. This is a rod which sticks up vertically out of the mine. When it is bent or broken by the hull of the tank or its tracks, it causes the mine to detonate after a short period of delay. This period of delay allows the tank to proceed still further over the mine.

Another type of fuze is the horizontal tentacles fuze which consists of four fine wires or hoses which spread out radially from the mine fuze, at approximately right angles to each other. When two wires or hoses on opposite sides are crossed by the tank tracks simultaneously, the mine detonates.

Influence fuzes operate on a target signature and not by physical contact or direct pressure. Influence fuzes can be designed to operate on the thermal, seismic, acoustic or magnetic signature of the target, or by a reflected radar or laser beam, or by combination of more than one such signature.

Fuzes for off-route mines can be operated in a variety of ways, although they are normally of the simple pull/trip wire, frangible wire or sensitive wire type. Other more sophisticated types include those that are operated by electric actuators, or by interrupted infra-red beam actuators, or by a combination of all the features mentioned.

River Mine

The Dutch have developed a mine for use in rivers and canals. It comprises a sensor assembly and fuze which actuates the mine to which it is attached. The sensor assembly releases an actuator float or "fish", which detonates the mine beneath it when approached by a swimming, schnorkelling or wading tank.

ANTI-PERSONNEL MINES

Introduction

The role of anti-personnel mines is to kill or maim enemy soldiers on foot, although some mines are capable of disrupting wheeled vehicle tyres. Anti-personnel mines are used in a number of ways. In conjunction with an anti-tank mine field, to prevent crossing, reconnaissance or hand breaching by soldiers on foot, and to attack tank crews dismounting from damaged vehicles. Anti-personnel mines are also used to deny the enemy the use of well-defined routes or particular areas where it is estimated that soldiers would normally travel on foot, such as narrow routes in urban or forest areas. They are used for local defence in the form of protective mine fields around prepared infantry positions. They are particularly effective in the harassing role when isolated mines or groups of mines are left in random or attractive areas. Unless these mines are carefully recorded and controlled, however, they can prove extremely hazardous to friendly troops in event of counter-attack, or to the civilian population.

CLASSIFICATION OF ANTI-PERSONNEL MINES

Individual Attackers

Individual attack mines are designed only to attack one man. Their area of effect is limited; ideally they should maim the target man by shattering his foot, but their effect is not always easy to predict. Individual attack mines are classified as either "ground emplaced" or "scatterable" anti-personnel mines. A ground emplaced mine is buried or positioned as part of a deliberate laying drill. Laying is a slow and tedious business, but the positions and numbers of mines can be closely controlled and recorded. Figure 3 shows typical examples of this type of anti-personnel mine, the British Elsie mine and the Dingbat mine. Elsie consists of a small plastic case which is pressed into the ground by foot pressure. A small shaped charge is then inserted and the mine is live. Dingbat is a blast mine about the size of a boot-polish tin. Scatterable mines are designed to be

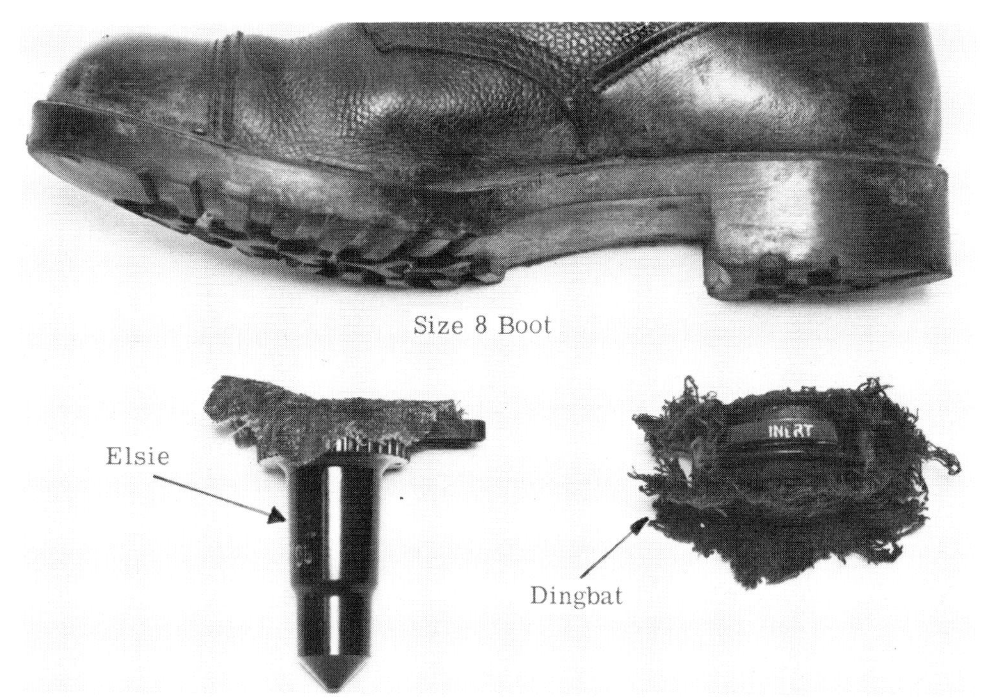

Fig. 3. Typical ground emplaced anti-personnel mines

rapidly laid either by a moving vehicle or helicopter. They are not usually buried achieving concealment by their small size, or by camouflage scrim attached to their upper and lower surfaces. Generally, they are of the blast type, since it cannot be guaranteed that they will land the right way up. Their size makes them difficult to detect on the ground. Typical of the scatterable individual attackers is the British Ranger mine which is designed to be remotely delivered at random into an area 60-100 metres from the launch assembly.

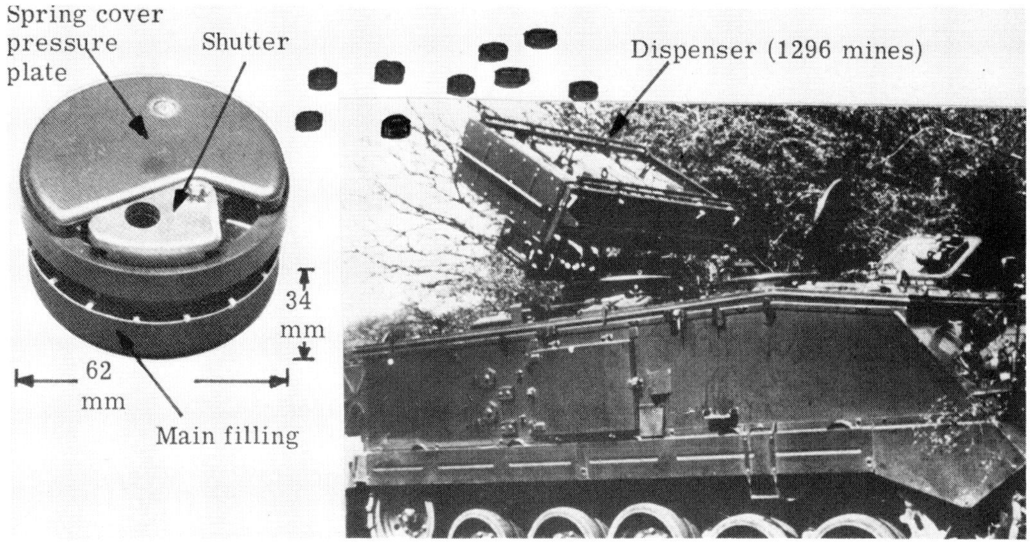

Fig. 4. The Ranger anti-personnel scatterable mine and dispenser

Mass Attackers

Mass attackers are larger anti-personnel mines which achieve their effect by normal blast attack, the blast propelling pre-formed fragments such as steel balls or chopped steel bar, over distances up to 100 metres. Mass attackers have either a random all round effect or a directional effect. Random all round effect mines are usually detonated at some distance above the ground. They are either mounted on stakes, or designed with the assembly buried, the central core being blown into the air by a cordite charge and detonated by a timer or fixed wire when the mine has reached about one metre above ground level. This sort of mine is normally actuated by a foot switch or trip wire. Alternatively it is possible to design mines so that their load of pre-formed fragments can be propelled in one selective direction. The American M18 Claymore mine, which blasts 700 steel balls out over a 90° arc with a lethal radius of 50 metres is an example of this type of directional anti-personnel mine. Directional effect mines are either trip wire operated or command-detonated by an operator.

SUMMARY

Anti-personnel mines have significant disadvantages. They can be hazardous to lay, and are very difficult and dangerous to clear. They create a hazard to the civil population, which can remain for many years, unless a reliable self-destruct or self-sterilising element is contained in the fuze. This is a feature, however, which is missing from almost all anti-personnel mine designs. Anti-personnel minefields in Korea caused more casualties to own troops than they did to the enemy. Mines laid by the Australian forces in Vietnam were removed by the Vietcong, and used against the Americans, South Vietnamese and Australians. Anti-personnel mines, and indeed mines in general, must therefore be viewed as potentially two-edged weapons, and used only under strictly controlled conditions, with areas of their use being carefully recorded and promulgated.

METHODS OF LAYING MINEFIELDS

Mines can be laid in a number of ways, by hand, by mechanical means, surface laid from a moving vehicle, by helicopter, or remotely delivered from a gun, rocket, mortar or dispenser tube.

Hand Laying of Minefields

The hand laying of minefields is the traditional method of mine laying, but it is slow and tedious. It is, however, the best way to camouflage the mines successfully and is often used when nuisance mining or in the creation of small protective minefields. It is also necessary in places where mines or their fuzes cannot be mechanically laid. Hand laying is time consuming and this must always be borne in mind when considering this method.

Mechanical Mine Laying

Mines can be laid below ground level by mechanical mine layer. To do this a mechanical mine layer requires: an agricultural disc to cut the surface of the soil, a double sided plough to lift the earth, a device which arms the mine fuze (this can be a team of men), a chute to deposit the armed mine in the furrow at the required spacing, and smoothers to replace the earth.

Surface Laying

Mines can be surface laid both from a mechanical mine layer and from an ordinary moving vehicle supplied with some form of chute. This is a quick alternative to buried mining, and provided that the mines are intelligently sited, they are difficult to detect from a closed down tank at night. Mines can possibly be dug in later if time permits. There are, however, some disadvantages with surface laid mines. Double impulse fuzes will not actuate under the second or subsequent road wheels of a basic tank, because without the natural elasticity of the dug ground to help, the diaphragm spring in the fuze does not return to its first position due to the tension of the track; the second pressure is therefore not taken up unless a subsequent vehicle drives over the same mine. Plastic mines can be physically broken by the tank tracks.

Alternatively, mines can be surface laid from a helicopter. Although this method looks attractive, a helicopter has only a limited capacity to carry mines, and unless the helicopter moves very slowly there are large gaps between the mines as they cannot be loaded into the dispensing chute quickly enough. The helicopter, therefore, is very vulnerable when it is surface laying mines, as it has to travel virtually at walking pace and at ground level.

Remotely Deliverable Mines (RDMs)

Remotely deliverable mines are small mines that can be delivered by aircraft, artillery shell, rocket or short range mortar bomb. Mines launched by such means have to be small, to fit into the carrier shell or rocket. They also have to be very robust to counter the high acceleration loads imposed during firing or launch, and the impact stresses when they come to earth. One such example is the US M718 155 mm minelet round which carries nine belly attack kinetic energy mines of low weight.

MINE COUNTERMEASURES

Introduction

Mine countermeasures include any action whose purpose is to neutralise, destroy or disrupt a mine, or render it inactive. This can be done by the physical removal of the mines by hand (hand lifting) or by mechanical means. Alternatively, the mines can be detonated by indestructible devices, explosives, or in the case of influence fuzed mines, by electronic means. They can also be countered by shielding the mines by earth or foam.

Hand Lifting

Hand lifting is the traditional method of mine clearance, and it is perhaps the only reliable way of removing and neutralising non-metallic mines. It is, however, a slow and dangerous procedure which has been made more so by the small size and easy concealment of modern non-metallic anti-personnel mines. The process of mine clearance is divided between detection and removal.

Detection

When minefields are cleared by hand, detection is either by hand, prodding with a prodder or bayonet, or by mine detector. Mine prodding is slow but relatively safe, although a number of fuzes can be employed that are sensitive to metallic prodders, or actuate on movement of the mine. Current mine detectors work on a principle of low frequency (LF) induction and electrical impedance, and therefore only pick up metallic mines or mines with a high proportion of metallic components in their construction. This type of detector is useless against fully non-metallic mines, and indeed can actuate some kinds of influence fuzes.

Removal

Mines when found are marked and by-passed, for "pulling" or destruction later. Mine fuzes may well incorporate anti-handling switches, which render this task very hazardous. Again the point must be made that this is a time consuming task. For example, working with the current range of mine detectors and mine prodders, one Sapper troop of 30 men can breach a lane 120 metres by 8 metres in 6 hours by night with no moon.

Mechanical Removal

The simplest form of mechanical removal is the plough. Ploughs skim off the earth in front of the tank tracks and roll it away to one side with the mines that it contains. Current ploughs in service are the Soviet KMT4 and KMT6. These ploughs employ tines which comb through the ground and bring the mines to the surface, where an angled blade pushes them beyond the area of the tank tracks. The KMT4 and KMT6 are widely deployed in the Warsaw Pact armies.

The fitting of ploughs to standard gun tanks, however, can impose a considerable strain on the steering and transmission clutches of the tank. This strain is made worse by stony soil or roots, which get tangled up in the plough tines and force the plough to the surface or stop the vehicle. Furthermore, the ploughs can be defeated by deep buried mines, laid so that the ploughs pass over them without disturbing them, or by mines fitted with anti-disturbance fuzes, although these merely destroy the plough rather than the tank behind them. Plough tanks can also be defeated by influence fuzed belly attack mines which pass between the mine ploughs, and by tilt fuzed mines. A tilt fuzed mine can be countered by fitting a chain or grid across the front of the tank to detonate the mine before the belly of the tank can pass over it.

DETONATION OF MINES BY INDESTRUCTIBLE DEVICES

The most basic form of mine clearance is to detonate the mine by impact or crushing with a heavy weight. This can be done by flails, rollers or indestructible vehicles.

Flail vehicles originated in 1941 with the "Baron" based on the Matilda tank, and the best known was the "Crab" flail, based on the Sherman tank. The flail consisted of a rotor, to which were attached 43 weights on a steel wire rope or chain. This was carried forward of the vehicle and rotated at high speed by a power take-off from the engine. The vehicle travelled at $1\frac{1}{2}$ mph when flailing. This method of mine clearance was not 100% effective, and mine clearance devices of this kind can be defeated by double impulse fuzes.

Rollers have ranged from small $\frac{1}{4}$ ton rollers, a number of which are carried on a train forward of the vehicle, to a single 8 ft diameter roller weighing 31 tons. The best known rollers are those which are in current use with the Warsaw Pact armies. These were first fitted to the T54 and T55 tanks and are currently used in conjunction with mine ploughs on T62 tanks. The disadvantages of rollers are that they are heavy and cumbersome, reducing the cross country performance of the tank very severely. They are also defeated by double impulse mine fuzes. They are, however, relatively indestructible and require no special training of the crew, and are often used to detect the edge of a minefield by detonating a mine in the first row. The mine ploughs can then be deployed to clear the rest of the minefield.

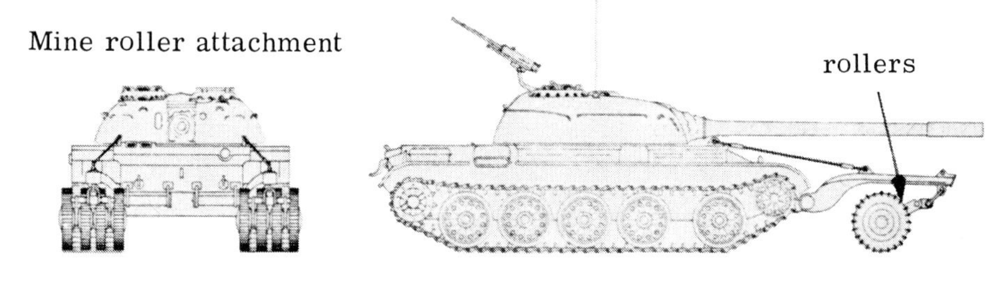

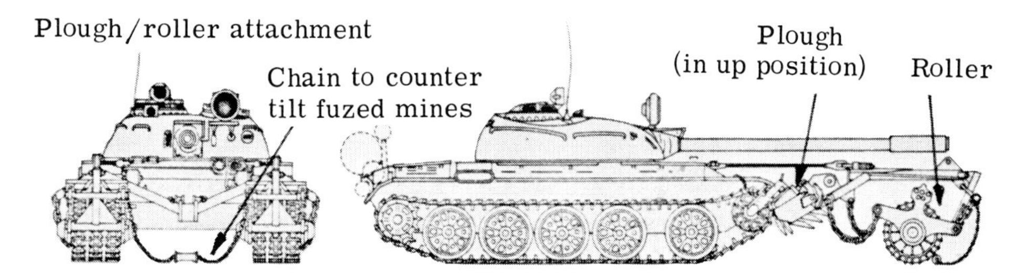

Fig. 5. Plough and roller mine countermeasure equipment

Experimental work on specialised tracked vehicles resistant to mine blast has at various times been undertaken in the UK and US. At present this is probably the only method whereby all types of anti-tank mines can be found and detonated at a speed acceptable to the service, however, no indestructible vehicle, either remotely controlled or human operated, has yet reached service in any army in the world.

DETONATION OR DISRUPTION OF MINES BY EXPLOSIVES

Rocket Hoses

Most kinds of mines are susceptible to activation or physical disruption by the detonation of explosives near them. One method of achieving this is to propel, by rocket, an explosive hose across the minefield. This explosive hose is then detonated, and the blast clears a path through the minefield. This method of mine clearance is done by the UK Giant Viper, which clears a gap 180 metres long by 8 metres wide in a minefield.

Fuel/Air Explosives

Minefields can also be disrupted by the use of fuel/air explosives to produce blast overpressures of sufficient magnitude to detonate the buried or surface laid mines. The fuel/air explosives are contained in mortars or rocket shells, canisters dropped by helicopters or propelled by rockets. In the case of the rocket method of projection, the vehicle carrying the rockets is deployed near the forward edge of the minefield, and the rocket projectiles are then fired out into the minefield at intervals of 10 metres. The explosion produced by the fuel/air explosives in the rocket or mortar bombs is, in theory, enough to detonate any mines within a circle whose radius is 15-20 metres.

Unfortunately explosive clearance devices fired by a rocket are very susceptible to firing conditions. It is also difficult to align explosive hoses exactly in the direction required, the accuracy of alignment is degraded by side winds or other meteorological conditions. Certain types of fuzes are resistant to blast attack, since they require a longer pulse time than the very sharp pressure rise produced by blast detonation. No explosive method of mine clearance is 100% reliable, and so in practice these explosive clearance devices are used in conjunction with mine ploughs.

DETONATION OF INFLUENCE FUZED MINES BY ELECTRONIC MEANS

Mines that are actuated by influence fuzes can be subjected to countermeasures. Where fuzes are actuated by seismic or magnetic influence, seismic and magnetic target signatures can be produced artificially to actuate the mines ahead of the tanks they are designed to destroy. In addition, it is possible by the same kind of process as used to de-magnetise ships, to alter the magnetic signatures of tanks, to below the level to detonate the mine.

SHIELDING MINES BY EARTH OR FOAM

Mines can also be shielded by covering them with earth or foam, so that vehicles passing over them will not actuate the fuze mechanism. This can be done by heaping large quantities of earth onto suspected minefield areas by bulldozer. This is a noisy, time consuming task and is really only a starter for the clearance or neutralisation of rear area minefields. Alternatively, suspected minefield areas can be covered with expanded polyurethane foam. This foam not only blankets the mines against downward pressure from tanks, and in the case of anti-personnel mines from foot pressure, but it also if raised to sufficient height above ground level locks tilt fuze masts in the vertically upright position. It does not, however, prevent frangible masts from actuating the mines beneath them. The disadvantages of this method of mine neutralisation is the large amount of chemicals required to produce the foam, and the difficulty of effecting such a foam minefield crossing under observed fire or assault conditions. At present the foams available for such use have a significant setting and curing period, which reduces the speed at which the foam generating vehicles can be moved forward.

MINEFIELDS

Mines, especially anti-tank mines, are rarely used on their own. They are most effectively employed in minefields to filter, delay and deflect enemy armoured thrusts. Thus, although a good destroyer of tanks, the most disruptive characteristic and benefit to own troops of a minefield is the fact that it canalises enemy armour into areas where such armour can be easily engaged by direct fire weapons. Tank columns traversing minefields even when led by tanks fitted with mine ploughs, are confined to the speed of the plough tank, about 4-8 kph, and are thus easy targets for anti-tank fire. This of course pre-supposes that minefields are in all cases covered by aimed direct anti-armour fire.

In theory, the design of minefields depends on the effectiveness - the "stopping power" - required at the site of any particular minefield. This stopping power will depend on the type of mine and fuze, and the number of rows laid, which in turn will depend on the numbers of mines available, the time available and the means of laying available.

Classification of Minefields

For planning purposes, minefields are sub-divided into types and known as Tactical, Protective, Phoney and Nuisance. Tactical minefields deflect the enemy into required killing areas, and they deny ground to the enemy for limited periods. Protective minefields are small minefields laid by non-engineers units to give themselves local protection only. Phoney minefields, as the name implies, are minefields which are marked and fenced as minefields, but contain no mines. The remaining type, the Nuisance minefield is a small minefield designed to delay and disorganise enemy forces and to lower their morale. This is done by mining in depth along their likely main axis during the withdrawal of own troops. Sometimes the term "minimal minefield" is used. A minimal minefield is a hastily laid minefield of a few rows, possibly surface laid because of the lack of time or resources to lay a tactical minefield.

Recording and Marking of Minefields

For very obvious reasons, the laying, recording and marking of minefields is an absolute necessity and must be very strictly controlled, at the highest practicable levels of command. It is essential that the siting of a minefield is cleared in advance of laying, to prevent jeopardising future plans and intentions. When laid, a detailed minefield record must be maintained. This record must be kept in considerable detail showing the location of the start and end of mine rows, the numbers of mines and the use of anti-personnel mines and anti-handling devices. Minefields are required to be marked with a perimeter fence of double strand wire at least 1.25 metres high on the rear and both sides. These fences have to be marked with red triangular minefield markers. The front, enemy side, of the minefield is also marked with a fence, but it is single strand, knee high and does not have marking signs. Minefield fences do not have to conform exactly with the shape of the minefields they surround, for obvious tactical reasons.

SUMMARY

Mines are used to thicken up existing obstacles, and to channel and deflect the enemy into chosen areas where he can be engaged in circumstances more favourable to the defender. Mines, however, can be double edged weapons, and once laid cannot just be forgotten. Although minefields can be designed to defeat countermeasures, they have limitations, and are expensive in time, manpower and resources to lay.

SELF TEST QUESTIONS

QUESTION 1 What are the main component parts of a mine?

Answer ....

....

....

....

QUESTION 2 What are the required characteristics of a mine?

Answer ....

....

....

....

QUESTION 3 What are the main features of the tank that the anti-tank mine is designed to attack?

Answer ....

....

....

....

QUESTION 4 Name the types of anti-tank fuze.

Answer ....

....

....

....

QUESTION 5 How are anti-personnel mines classified?

Answer ....

....

....

. . . .

QUESTION 6 What are the disadvantages of anti-personnel mines?

Answer . . . .

. . . .

. . . .

. . . .

QUESTION 7 What are the main methods of laying minefields?

Answer . . . .

. . . .

. . . .

. . . .

QUESTION 8 How does the anti-tank plough work and what are its disadvantages?

Answer . . . .

. . . .

. . . .

. . . .

QUESTION 9 Describe briefly the current ways of minefield clearance by explosives.

Answer . . . .

. . . .

. . . .

. . . .

QUESTION 10 How are minefields classified? Give a brief description of each type.

Answer . . . .

. . . .

ANSWERS ON PAGE 260

16

Pyrotechnics

INTRODUCTION

Pyrotechnic comes from the Greek Pyr (fire) and Techny (art) and covers the family of compositions designed to give special effects such as noise, light, smoke etc. It was, for long, more of an art than a science, being well known on November 5th in the form of fireworks. Modern military pyrotechnics stem from the work done in the 19th century by men such as Congreve and Boxer who designed large rockets with reasonably reliable performance. In World War I, the combination of trench warfare and the advent of the aeroplane led to a further demand for pyrotechnic devices of all kinds for signalling and illumination. Since then a wide variety of items have been developed and introduced into service. In the last decade a great resurgence in the pyrotechnic field has been apparent. This has been fostered, partly by the exacting needs of space exploration and partly by the most stringent requirements for specialist military uses of pyrotechnics with the emphasis on reliability, long shelf life and optimum performance.

SCOPE

The term pyrotechnics covers illuminants, smoke, incendiary, signal, tracer, delay and priming compositions. It also includes devices used for simulation and often involves the combination of two or more of the above effects.

CHEMICAL BASIS OF PYROTECHNIC COMPOSITIONS

The basic compositions in pyrotechnics consist of intimate mixtures of solid fuels and oxidants together with various additives as required to ensure burning at a uniform and predictable rate. Fuels are necessary to provide combustion and in some cases intense luminosity. Examples are magnesium, aluminium and boron. Oxidants provide the oxygen necessary for combustion of the fuel such as chlorates, nitrates and peroxides. The choice of an oxidant is sometimes dictated by a required colour of flame. In addition there are binders which improve cohesion of powdered ingredients; protect metal powders from corrosion; reduce the sensitiveness of compositions to impact and shock and sometimes serve as all or part of the fuel component. Examples of binders are waxes, resins, oils and varnish, but binders are not always necessary particularly where a device is designed to burn very quickly. Moderants may also be found where the rate of burning needs to be retarded and where temperatures have to be reduced.

ILLUMINATING COMPOSITIONS

Illuminating compositions are required to illuminate an area for the purpose of photographic or visual reconnaissance. The two main categories are rapid photographic flash and a long burning flare for visual purposes.

Photoflashes

Photoflashes provide the necessary light for aerial photography at night and are ejected from aircraft in the form of a cartridge or a bomb. They are required to have very high intensity, short burning time and a very short but predictable build up to peak intensity. They are mainly based on aluminium or magnesium powder as the fuel and nitrates of sodium, potassium, barium and strontium, and potassium perchlorate as the oxidant.

Illuminating Flares

Illuminating flares can be projected in a variety of projectiles, dropped from aircraft and helicopters or function near ground level. They may or may not be fitted with parachutes and have high intensity and long burning times. They have similar ingredients to photoflashes but the linear burning rate is much slower. An example of a typical flare is shown in Fig. 1.

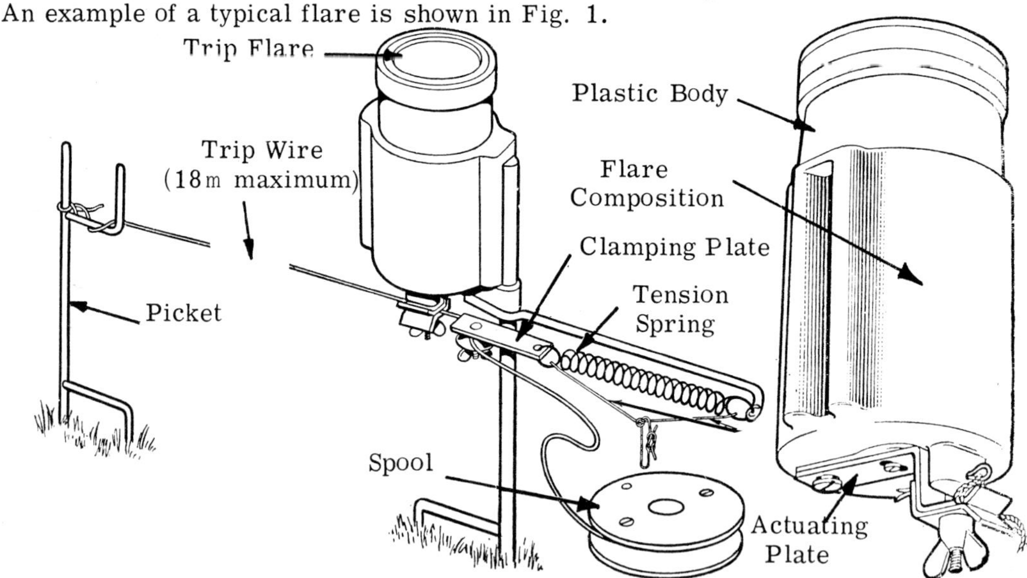

Fig. 1. Trip wire flare

SMOKE COMPOSITIONS

These are required to produce a dense smoke cloud either of a colour identifiable by day as a signal or as a whitish cloud for screening areas from observation.

Coloured Smokes

Coloured smokes, usually red, blue, green, orange and yellow, vary in size from a 1 inch signal cartridge to large artillery projectiles. The smoke is disseminated either as a puff or a large burst from a projectile or at a steady rate from a grenade. All coloured smoke employs a finely powdered dye stuff which is volatilised by other ingredients. These smoke mixtures must burn smoothly with moderate heat sufficient to vaporise the dye stuff without decomposing it. See Fig. 2 for an example.

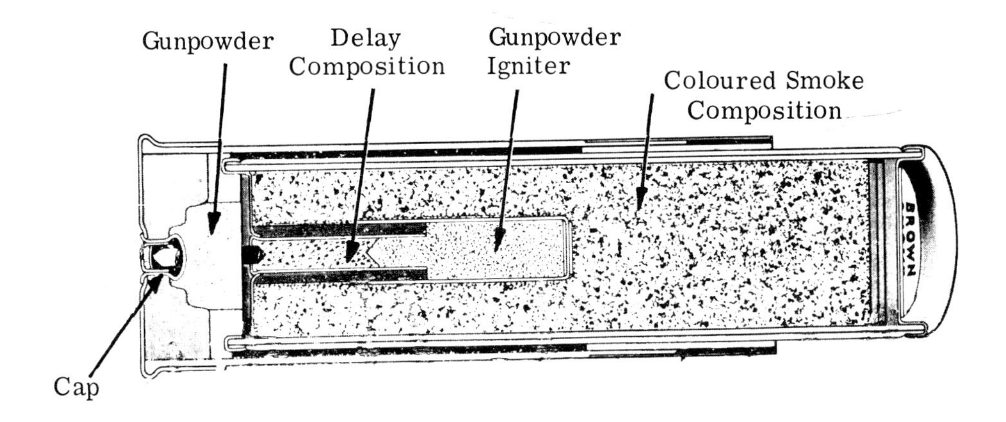

Fig. 2. Coloured smoke cartridge

Screening Smokes

Screening smokes are designed to screen own troops or vehicles etc from the enemy for short periods. They are usually whitish or grey aerosol clouds produced by chemical or pyrotechnic means. Chemical clouds are produced by rapid hydrolysis of certain liquids such as titanium tetrachloride and are simple to use. Such liquids however are unpleasant to handle. Pyrotechnic clouds use white phosphorus which burns spontaneously in air or red phosphorus which needs high explosive to assist its effects and these are both referred to in the service as bursting smoke when filled into a projectile. Another well used type uses hexachlorethane which provides a more consistent though slower smoke build up. The performance of smoke depends partly on the moisture in the atmosphere because the reaction produces minute highly reflecting particles or droplets that scatter light.

INCENDIARY COMPOSITIONS

Incendiary compositions have been used in various forms from very early times. Catapults were used for projecting inflammable mixtures of pitch, sulphur etc to ignite wooden structures sheltering the enemy. Greek fire is referred to in many war books and was said to have been used around 700 AD. Modern incendiary compositions came into being during World War I and received a further expansion in World War II. Incendiary materials must have a high heat of reaction and the heat must not be released or liberated too quickly. There are two basic groups of incendiary compositions; metals which burn in the air once ignited, and combustible fluids that spread over the target while burning. Aluminium and magnesium are typical examples of the former whilst petrol and napalm are included in the latter type.

SIGNAL COMPOSITIONS

Signal compositions are usually coloured and are used to convey messages or information over various distances. Such signals could be simple warnings, recognition codes, markers and so on. In each case they must be readily identifiable and have a clear and consistent colour output regardless of battle and atmospheric conditions. They are usually self luminous compositions and burn with a coloured flame. For service use they must have a known and consistent burning rate and have sufficient intensity over the specified range and time. The colours which are most readily identifiable are red, yellow, green and white. Strontium is used for red, sodium for yellow, barium and potassium for green and their nitrates are used for white signals. Various dyes are also incorporated. An example of a signal cartridge is shown in Fig. 3.

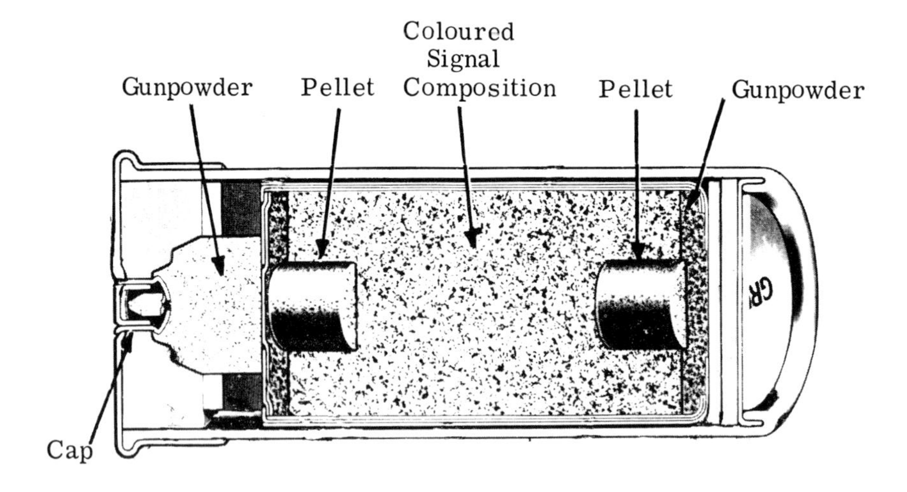

Fig. 3. Signal cartridge

TRACER COMPOSITIONS

Tracer compositions are used in projectiles to show the trajectory and line of flight. They must be capable of operating efficiently under high rates of spin and survive high firing stresses. The essential ingredients are magnesium plus an oxidising agent such as a nitrate, perchlorate or peroxide. Quality of colour can be improved by the presence of chlorine and the magnesium must be protected from oxidation by such substances as wax, shellac resin etc. which may also act as binders. Some tracers have a dark ignition phase which prevents the light emitted from blinding the firer - after the required period the tracer burns normally. An example of a tracer arrangement is shown in Fig. 4.

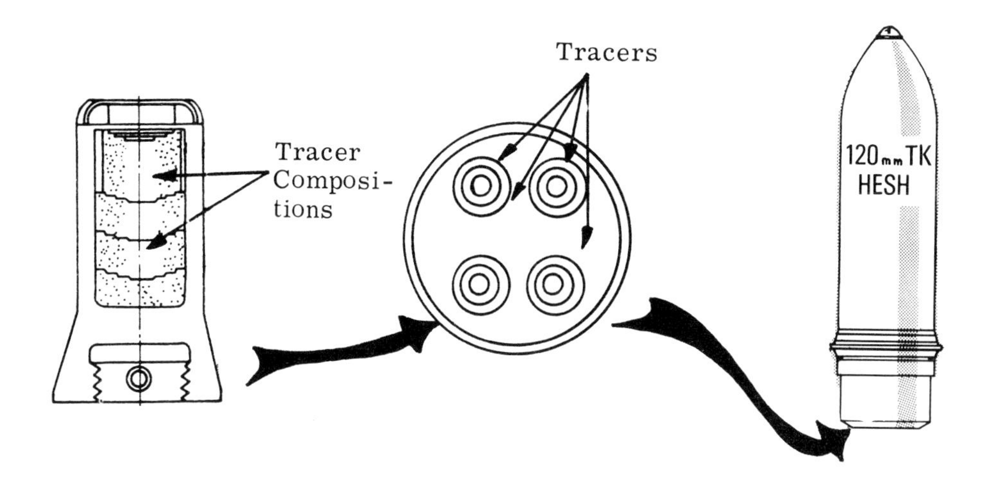

Fig. 4. Base tracers in a tank practice round

DELAY COMPOSITIONS

A pyrotechnic delay is a device which makes use of the time taken to burn a train of inflammable composition to provide an interval of time between events. The simplest form is a pressed delay pellet which burns through in a known time or a length of safety fuze burning at a slow rate. These are not accurate and due allowance must be given for the tolerance inherent in the design and method of filling. More accurate types are available such as explosive delay motors in some ballistic missiles. Delays can operate over a wide range of times from a few milliseconds to several minutes depending on the requirement.

Two basic types exist namely gas producing and gasless. The gas producing types usually contain organic matter and on combustion produce large volumes of gas. The rate at which they burn depends on the ambient pressure, and the gas produced must therefore be able to vent to the atmosphere if the burning rate is to be consistent. Gasless types, where the materials used produce mainly solid residue with little or no gas, are much less affected by ambient pressure conditions and can therefore be burned in confined spaces. The main constituents are generally a metal powder and a metallic oxide which interact by oxidation reduction which generates much heat. Examples of delays are shown in Fig. 5.

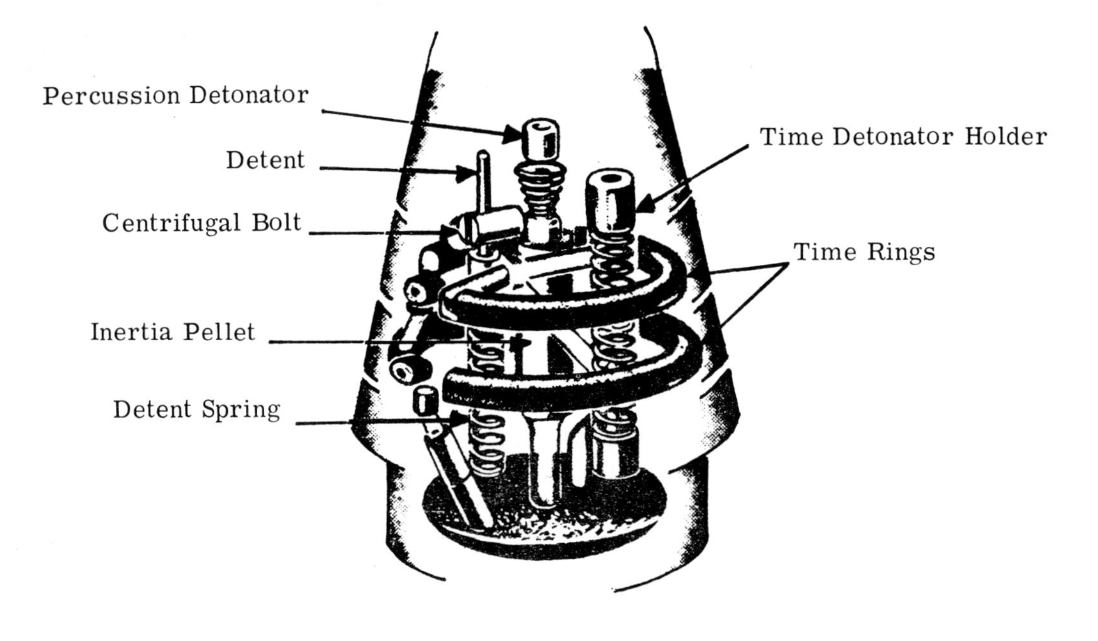

Fig. 5. Fuze showing time rings

PRIMING COMPOSITIONS

Pyrotechnic compositions are initiated by a flame produced from a variety of sources. The ease with which this takes place depends on the ignition temperature of the material used together with the physical condition of its surface and the characteristics of the flame itself. Mixtures of chlorates and organic materials ignite in the range of 150-250°C, starches and resins when finely divided need temperatures of 250-350°C whilst mixtures of metal powders and nitrates have ignition temperatures of 500-600°C. Ignition is a temperature effect and a time interval is needed to raise the temperature of the pyrotechnic surface to its ignition point. This is achieved by priming the composition to ensure that sufficient heat is brought to the main filling for the required period of time.

Characteristics of priming compositions are: they should be easily ignited by flame or flash, generate much heat, produce some solid products without burning violently and be compatible with the mixture with which they are used. Although easily ignited they should not be unduly sensitive to impact or shock.

A typical composition contains potassium nitrate, powdered silicon and sulphurless mealed gunpowder.

SIMULATORS

It is necessary to mention simulators under pyrotechnics although their usage is rather different from the main family. Simulators are required for two main purposes, to simulate the bursting of various ammunition items and to simulate the operation or firing of various weapons. There is a wide variety of simulators in service use and some are used in a training environment whilst others are operational stores.

Simulators of rifle and machine gun fire can be dropped from aircraft into the battle area. They consist of various components mounted on a base board and secured by clips, battens or wires.

Simulators of mortar fire use sound units initiated by a firing unit mounted in a box. These can also be dropped by aircraft.

There are even simulators of signal pistols for dropping from aircraft.

Simulators of gunfire and gunflash consist of a papier mache container fitted with a detonator tube and filled with flash composition. Some of these simulators have a double walled vessel containing a bursting charge surrounded by a mixture of acetone and aviation spirit which produces the flash effect. An example is shown in Fig. 6.

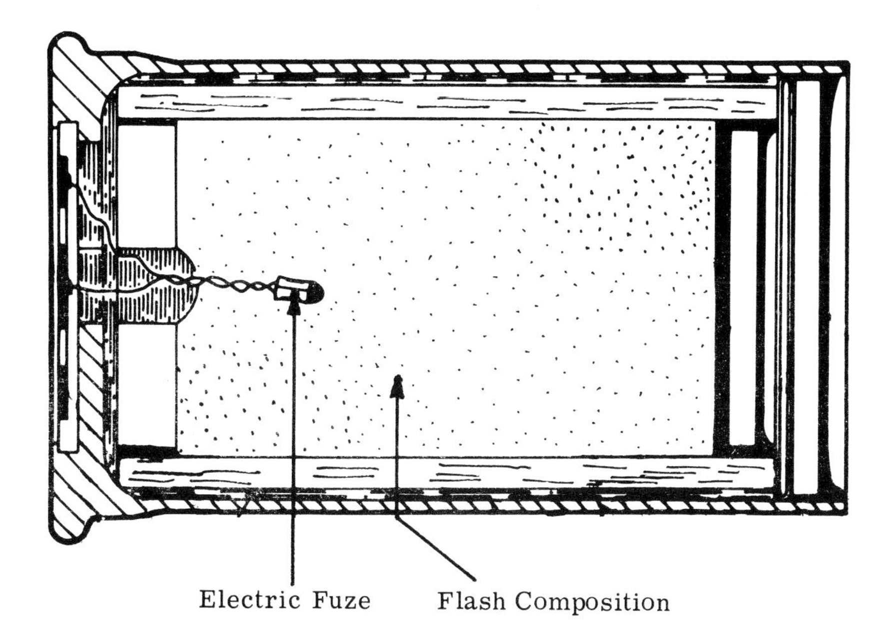

Fig. 6. A gun fire flash and sound simulator

Thunderflashes were introduced as a training store to simulate battle noises such as shell grenade or bomb bursts. They consist of a rolled paper cylinder containing flash composition and igniter. They have detached strikers which ignite the igniters by friction. An example is at Fig. 7.

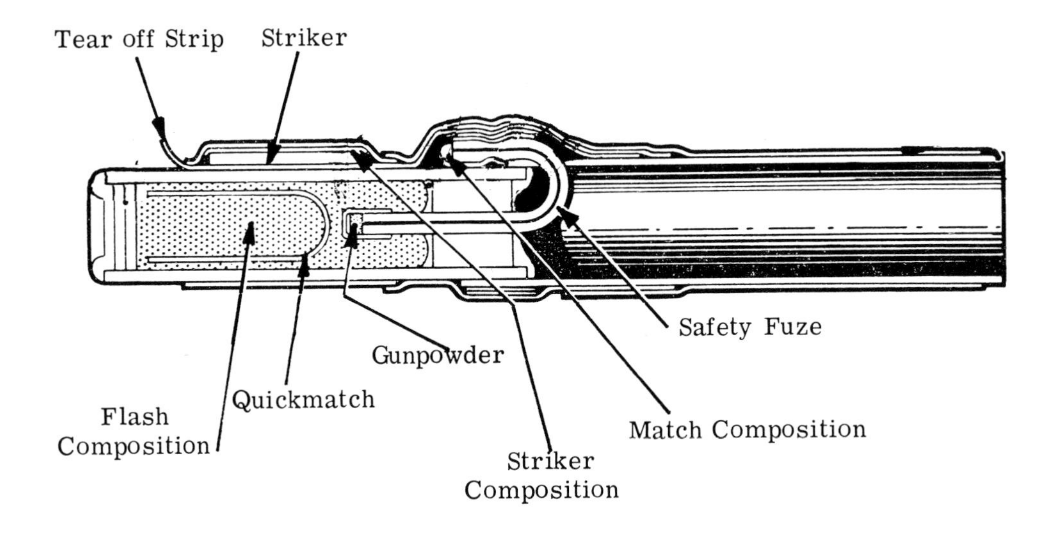

Fig. 7. Thunderflash

Over the last few years a system called "Simfire" has been introduced. It simulates the firing of a machine gun and the main armament of a tank and includes the process of aiming and firing. It also provides an indication of the fall of shot. The pyrotechnic components are simulator gunfire AFV flash and sound and simulator projectile burst orange smoke. A tank fitted with the complete equipment can operate as an attacker or a target.

CONCLUSION

The wide variety of pyrotechnics used in modern training and warfare, each item having its particular use and characteristics, will undoubtedly continue to play an important part in service life. The various types of pyrotechnic store are constantly being updated and improved.

SELF TEST QUESTIONS

QUESTION 1 What do you understand by the word pyrotechnic as applied to the services?

Answer ....

....

QUESTION 2 Why is it necessary to have a fuel and an oxidant?

Answer ....

....

QUESTION 3 What is a binder? Give examples.

Answer ....

....

QUESTION 4 Explain the difference between a photoflash and an illuminating flare.

Answer ....

....

....

QUESTION 5 For what are smoke compositions used?

Answer ....

QUESTION 6 What are the two basic groups of incendiary compositions?

Answer ....

....

QUESTION 7 Explain the use of signal compositions.

Answer ....

....

....

QUESTION 8 What ingredients are used to give red, yellow, green and white coloured effects?

Answer ....

QUESTION 9 Why are tracers required in projectiles?

Answer ....

QUESTION 10 Give a brief account of the use of simulators.

Answer ....

....

....

....

ANSWERS ON PAGE 262

17

Improved Conventional Munitions, Extended Range Projectiles and Terminally Guided Munitions

INTRODUCTION

The development of conventional gun fired ammunition for the future is being concentrated on making projectiles far more effective in terms of lethality, increased range and accuracy. To meet these requirements, new ranges of conventional ammunition natures have been coming into service since the late 1960s; the so called improved conventional munitions, (ICMs), extended range projectiles (ERPs) and terminally guided munitions (TGMs). Each type will be considered in this chapter.

IMPROVED CONVENTIONAL MUNITIONS

Improved conventional munitions are so called because they are designed to give an improved performance compared with standard HE shell against hard and semi-hard targets. Standard HE shell is designed specifically to attack people, but against modern mechanised armies there is also a requirement to have indirect fire ammunition which is effective against both tanks and armoured personnel carriers.

The performance of conventional ammunition can be improved against these types of targets in two ways, by placing a number of lethal sub munitions in a carrier type projectile or by placing a number of pre-formed fragments around an HE filling.

The first one we will consider is the carrier shell which ejects a number of sub munitions over the target. The main problem in achieving enhanced effectiveness compared with the standard HE shell in this instance, lies in resolving the dilemma between the quantity and the size of the sub munitions. There is a conflict between delivering sufficient numbers to give a satisfactory chance of achieving a hit, while at the same time having sub munitions that are large enough to do sufficient damage when they do hit.

The American M483 bomblet round (which is described more fully in Chapter 10 and see Fig. 13 of the same chapter) has 88 dual purpose armour defeating and anti-personnel grenades or bomblets. Each bomblet contains a HEAT warhead to attack armour and a pre-notched inside wall surface to provide controlled fragmentation effects against personnel. In fact, 24 of the bomblets are unnotched as they have to withstand the greater load felt by the rear of the projectile due to setback on firing. Carrier projectiles can also be used to remotely deliver small mines - "minelets" - as described in Chapter 15.

The other approach to improving conventional munitions is to use pre-formed fragments. The projectile shown in Fig. 1 is an example of this type of ammunition. It can be fitted with a proximity fuze and it is particularly effective against APCs. The pre-formed fragments, which are ball bearings in essence, are contained in a resin set around the explosive charge.

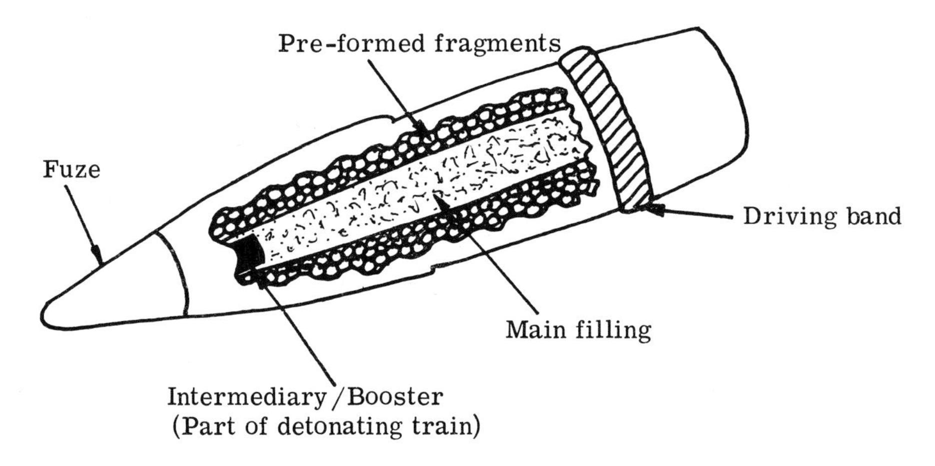

Fig. 1. A pre-formed fragment HE projectile

Thought is also being given to the use of various incendiary alloys as a means of enhancing warhead effectiveness. One such alloy, quasi alloy of zirconium, (QAZ) burns at 3500°F when subjected to detonation. It is an easily worked and cheap alloy and can be used to manufacture whole components such as fuzes, or it can be applied as a coating.

EXTENDED RANGE PROJECTILES

Introduction

No sooner is a weapon system introduced into service than increased range is demanded from it and its ammunition. This is particularly true of artillery weapons. The reasons for this requirement for extended range are many and various; not least the wish to engage and strike at the enemy and cause him casualties long before he can close with you. This is especially significant when faced with an enemy numerically superior in terms of men, material and equipment. Apart from designing a completely new weapon system, however, the options available to increase the range of an existing weapon system are threefold. The weapon's performance can be improved in a number of ways, or alternatively the performance of the ammunition can be improved by either improving and optimising it ballistically, or by providing it with some form of post firing boost while it is on its way to the target. The diagram at Fig. 2 summarises the three options which are looked at in some detail in the rest of the chapter.

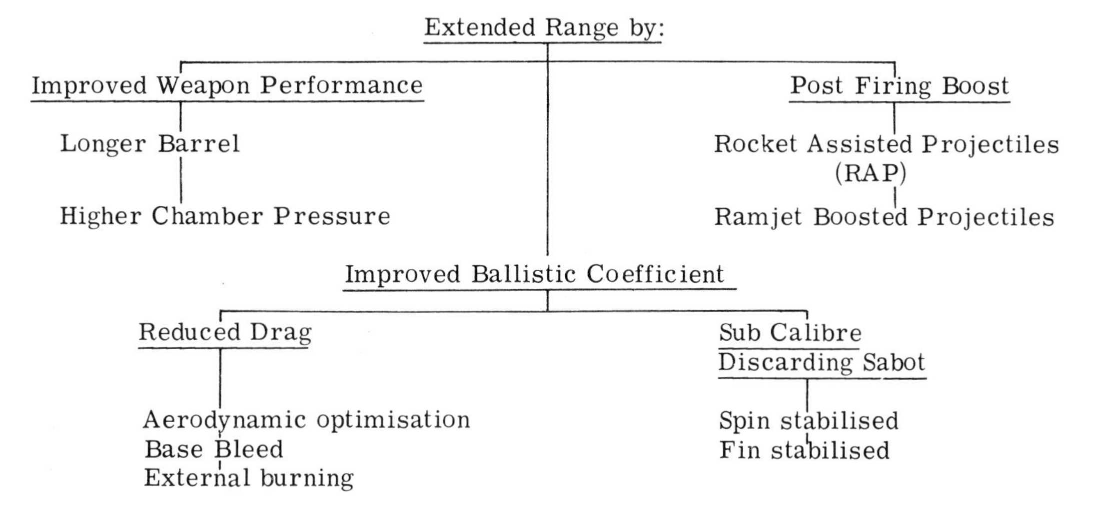

Fig. 2. Options for extending the range of projectiles

Improved Weapon Performance

In this instance all improvements are essentially aimed at increasing the muzzle velocity of the projectile, thereby increasing its range. To obtain higher muzzle velocities, however, the weapon must be modified, since increased muzzle velocities for a fixed mass projectile will require changes to the recoil system. Most weapon systems are designed so that their existing recoil systems can accommodate modest increases in muzzle momentum. Muzzle brakes can be added, or improved, to permit further small increases. Another method, the soft recoil system which is an American development, can be utilised. Soft recoil is a development by which the efficiency of a recoil system can be enhanced by employing stored energy which has to be overcome before the gun physically recoils. The main problem with the soft recoil system lies with overcoming the delay, fleeting though it is, in igniting the ammunition propellant charge. Standard propellant charges have an ignition delay of between 40 and 80 microseconds. Whilst this is not a problem with conventional guns, with soft recoil if the amount of run forward of the recoiling parts is too large before the gun fires, then the design problems involved in incorporating it into an existing gun system become too great. If the soft recoil system can be incorporated successfully, it can cut recoil by some 30%. This is a significant and worthwhile improvement for enabling increased muzzle velocities to be obtained without drastic modification to the gun. Soft Recoil is explained fully in Volume II, Chapter 5 of this "Battlefield Weapons Systems and Technology" series.

Increased muzzle velocities can be obtained also by fitting longer barrels to guns, but at the expense of making guns heavier with all the implications this has on their mobility, portability, size, shape and other such factors.

Alternatively, more propellant can be used to develop higher pressures with which to give the ammunition a bigger "kick" to project it from the gun. This will affect both the recoil system and the pressure within the chamber of the gun. Unless the soft recoil principle can be used, a larger and heavier recoil system will be required, and this, invariably, means a larger and heavier gun. Also, higher propellant pressures require a stronger chamber in the gun to contain them, and this again will probably result in a bigger gun. Furthermore, there is a "pressure break point" at which a further increase in propellant charge does not impart more energy to the projectile; the increased energy generated being expended in accelerating the increased mass of gas evolved. Finally, the projectile itself will require modification or redesign to strengthen it to withstand the increased firing stresses it will be subjected to.

Thus, there are limits to improving the performance of the gun as a means of obtaining extra range. Beyond a certain size, weight complexity and sophistication, a gun becomes impractical for field use, and ways and means of extending range have to be sought elsewhere. Although improvements to weapon performance would seem to have been exploited to practical limits, this is not the case with improving the performance of the projectile or ammunition itself to meet the extended range requirement.

Improved Ammunition Performance

As already stated, ammunition can be modified in two main ways to improve its range. Design can be concentrated on improving the ballistic coefficient of a projectile, or it can be designed to carry an energy source to give it an additional boost after it has left the gun.

Improved ballistic co-efficient. The ballistic coefficient of a projectile can be improved either by reducing its aerodynamic drag or by reducing its calibre.

Reduced drag. Reducing the coefficient of drag involves improving the ballistic shape, which in turn requires a basic redesign of the projectile. An example of an "aerodynamically optimised" projectile is shown at Fig. 2. The projectile as such does not have "walls" or parallel sides as the nose and tail are streamlined into each other. (Compare Fig. 3 with the conventional HE shell shown in Fig. 1 of Chapter 3).

The long streamlined ogive has to have lugs fitted to support it in the bore and to prevent the projectile from "side-slapping" its way up the barrel when the gun is fired. These lugs have themselves to be carefully designed to have good aerodynamic characteristics and, unless they are to be discarded on leaving the barrel, they have to be fitted to cant into the airflow to prevent loss of projectile velocity and hence range. If they are to be discarded, there are the additional problems of making them strong enough to withstand gun firing stresses and to remain attached to serve their purpose while in the gun, but easy to discard

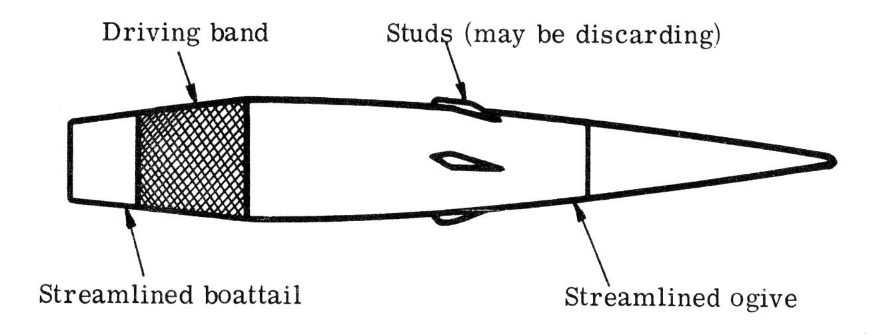

Fig. 3. An aerodynamically optimised projectile

shortly after leaving the muzzle. When discarding, the lugs should not be dangerous to own troops and cause casualties among them if the projectile is being fired over their heads. These problems apply equally to the driving band which should also, preferably, be discarded once the projectile has left the gun. This is necessary to reduce the turbulence of the air flow that occurs around the driving band when it remains fitted throughout the projectile's trajectory.

Projectile drag can also be reduced by streamlining the base of the projectile; "boattailing" the base as the method of streamlining this part of a shell is known. The most important disadvantage is the loss in payload. The reduced space available for the HE filling due to the streamlining, reduces the charge to weight ratio of the projectile by about 10% to 15%. So, although a 15% increase in range for a highly streamlined projectile is attainable, it is at the expense of a decreased payload and increased manufacturing costs (approximately 15% more than for conventional projectiles).

Another method of reducing drag that is used is known as "base bleed". This method, which was developed by the Swedes during the late 1960s, is shown in Fig. 4. It involves burning a quantity of propellant which is fitted into the rear

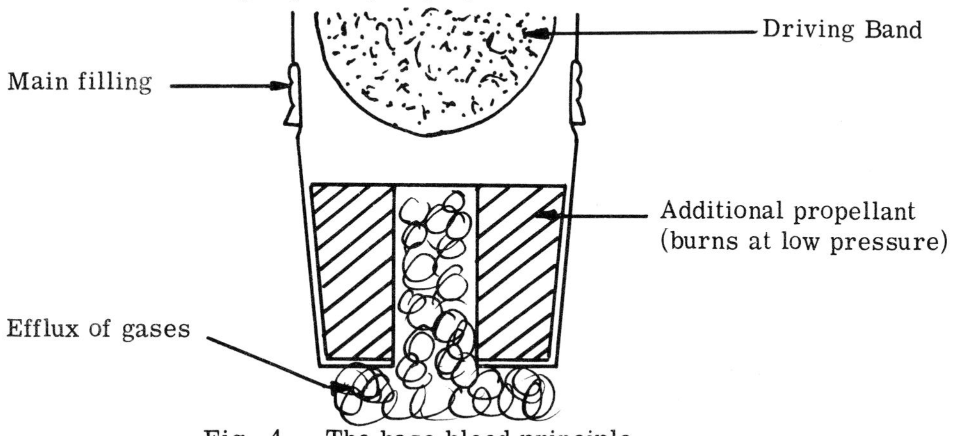

Fig. 4. The base bleed principle

part of the projectile. This propellant burns at low pressure and generates just sufficient gas to fill the partial vacuum that is always present at the base of projectiles whilst they are in flight. It is not designed to provide a jet or rocket effect at the base. Base drag, however, is reduced by up to 50%, giving range increases of between 10% and 20%. There is the attendant penalty to be paid in a reduced payload, but payload loss is not as great as it is with some of the other methods of increasing range discussed in this chapter.

The third method of reducing drag is by "external bleeding" which is really just a more sophisticated variation of the base bleed method. More propellant is used, and the gases generated are injected transversely through holes around the periphery of the base of the projectile into the boundary layer of recirculating, turbulent air flow at the rear. In theory, not only does this reduce drag by "smoothing out" the air flow, but it also provides a net thrust to the projectile. This technique is still (in 1981) in the early stages of development and is, as yet, not proven.

Sub calibre discarding sabot. Sub calibre discarding sabot projectiles as a means of increasing muzzle velocity and thus range have long been used for tank gun kinetic energy ammunition natures. APDS and APFSDS shot is covered in detail in Chapter 5. It is not a method much used by conventional artillery for the obvious reasons of serious loss in payload and the danger to own troops from the discarding sabot; a much greater possibility with artillery fired ammunition which is normally fired over own troops at the enemy. For large calibre guns, however, this method looks more promising. The calibre of the gun should be 175 mm or more to ensure that a sufficiently large sub projectile to accommodate a worthwhile payload is left to go on to the target after the sabot has been discarded.

Post firing boost. The other main area in which the performance of ammunition can be enhanced to give it additional range, is to provide it with a boost of some sort after it has left the gun. This can be done either by building a rocket motor into the projectile, or by designing it to utilise the ram jet principle.

Ram jet boosted projectiles. The original research and development into ram jet or air breathing projectiles was undertaken by the Germans in World War II. A ram jet projectile requires an inlet, a combustion chamber and a nozzle to be built into it; and shell with both peripheral and central ram jet systems have been considered. A ram jet projectile, however, is expensive, complex and the internal components required limit the volume of the explosive payload severely. A ram jet projectile has the potential for quite considerable range increases, in excess of 20%, and although requiring a degree of special handling, it does not require extensive modifications to be made to the conventional gun to be fired from it. This type of projectile, however, has not been introduced into general service use due to its high cost, complexity, and serious reduction in useful payload.

Rocket assisted projectiles. The concept of rocket assisted projectiles was also originated in Germany in World War II, although today it is being developed largely in America and Sweden. In a rocket assisted projectile a solid fuel rocket motor is incorporated into the rear part of the projectile, again at the expense of the explosive filling. Figure 5 shows the American M549 155 mm rocket assisted projectile to illustrate the basic design features of such a projectile.

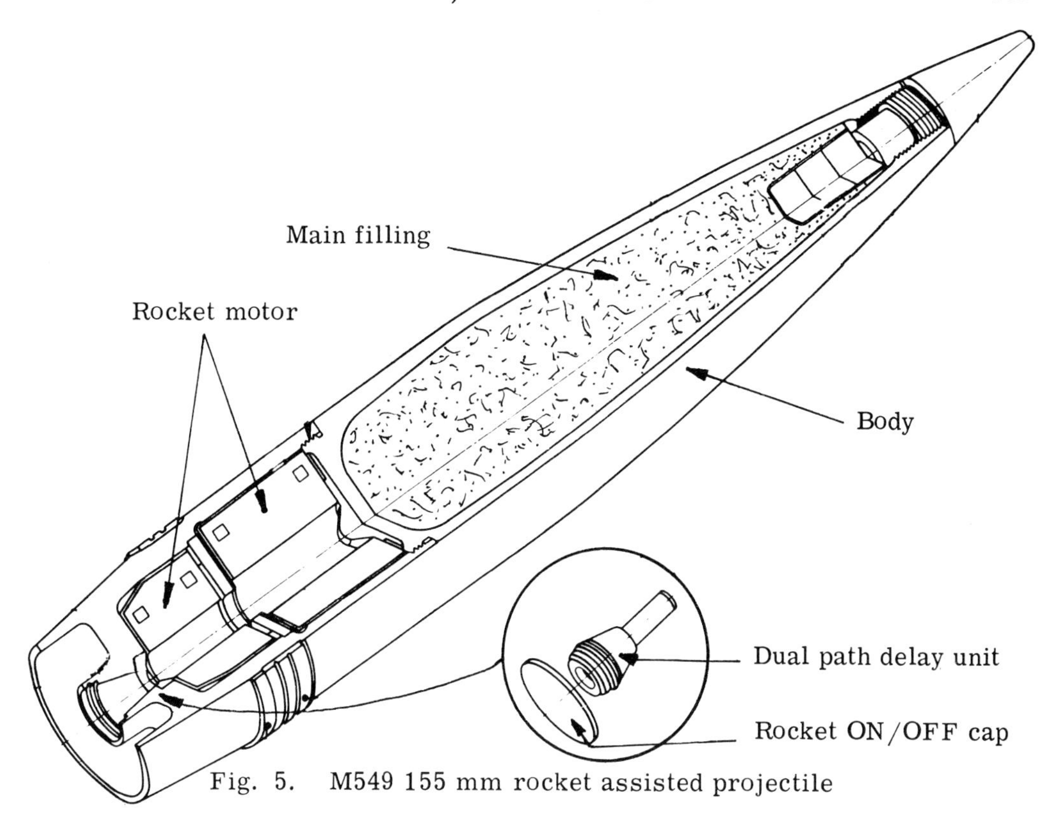

Fig. 5. M549 155 mm rocket assisted projectile

In theory, range is only limited by the quantity and type of rocket fuel incorporated into the projectile. In practice, the range is limited by a number of factors; the need to carry an effective payload, the need to match, ballistically, standard projectiles, the requirement for aerodynamic stability at trans-sonic speeds, and the need to accept the standard range of fuzes. Range increases of 20% are obtained using rocket assistance. For example the M549 round in Fig. 5 reaches out a further 6 kms over the standard 155 mm round to a range of 30 kms. Furthermore this type of projectile, being exactly similar externally to the standard projectile, does not require any special handling or modification to the gun. Nevertheless, there are a number of problems associated with rocket assisted projectiles. It is difficult to maintain the integrity of the rocket motor propellant when it is subjected to the considerable firing stresses in the gun. The motor propellant tends to distort and break up giving inconsistent burning which affects the accuracy of the projectile. Accuracy is also affected by the changing centre of gravity in the projectile due to the burning of the motor as it moves along its trajectory. The body of the projectile immediately around the motor has to be of a different grade steel to withstand the heat of the burning motor. This increases manufacturing costs and complexity and degrades fragmentation on the shell's dissolution of the target. Finally, there is the ever present drawback for all extended range projectiles, the loss of payload. Figure 6 is an extreme example, but it does show how much payload can be reduced to allow space for a rocket motor.

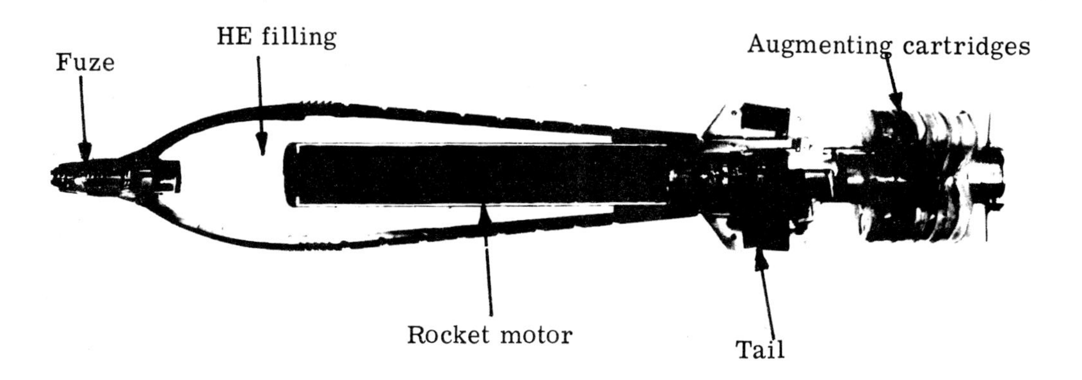

Fig. 6. 120 mm rocket assisted mortar bomb

Summary

Although it is technically possible to provide the extra range that is constantly demanded of conventional artillery weapons, there are a number of penalties to be paid for doing so. There are practical limits to achieving this increase by improved weapon performance, as guns become too big, barrels too long, stresses on chambers and recoil systems too great, rates of gun wear too high, design and manufacturing tolerances too demanding to be achieved other than at high cost, among many other factors. The alternative option of improving ammunition performance increases projectile costs as well as increasing the cost of target acquisition, as targets have to be acquired at the increased ranges. Useful payload and hence lethality is also reduced, as is accuracy since in a conventional gun accuracy is proportional to range.

TERMINALLY GUIDED MUNITIONS

The desire to engage the enemy further out with all the attendant problems involved in meeting this requirement, may be reduced if conventional artillery could be made more effective, that is achieve more impressive kill probabilities at existing ranges. At the present time it is estimated that it requires 1500 rounds of conventional HE shell to destroy one tank or 250 improved conventional munitions. The requirement is to kill a tank with only one or two rounds, and it is possible to achieve this with a projectile which can be terminally guided onto its target. The trend in development, therefore, for conventional artillery is towards the production of terminally guided munitions (TGMs) or precision guided munitions (PGMs) as they are also known, rather than for evermore elaborate and expensive guns and other types of "projectors". The American 155 mm Copperhead projectile, officially known as a Canon Launched Guided Projectile (CLGP), is the first such development in this trend in projectile design. A diagram of a typical TGM is shown in Fig. 7.

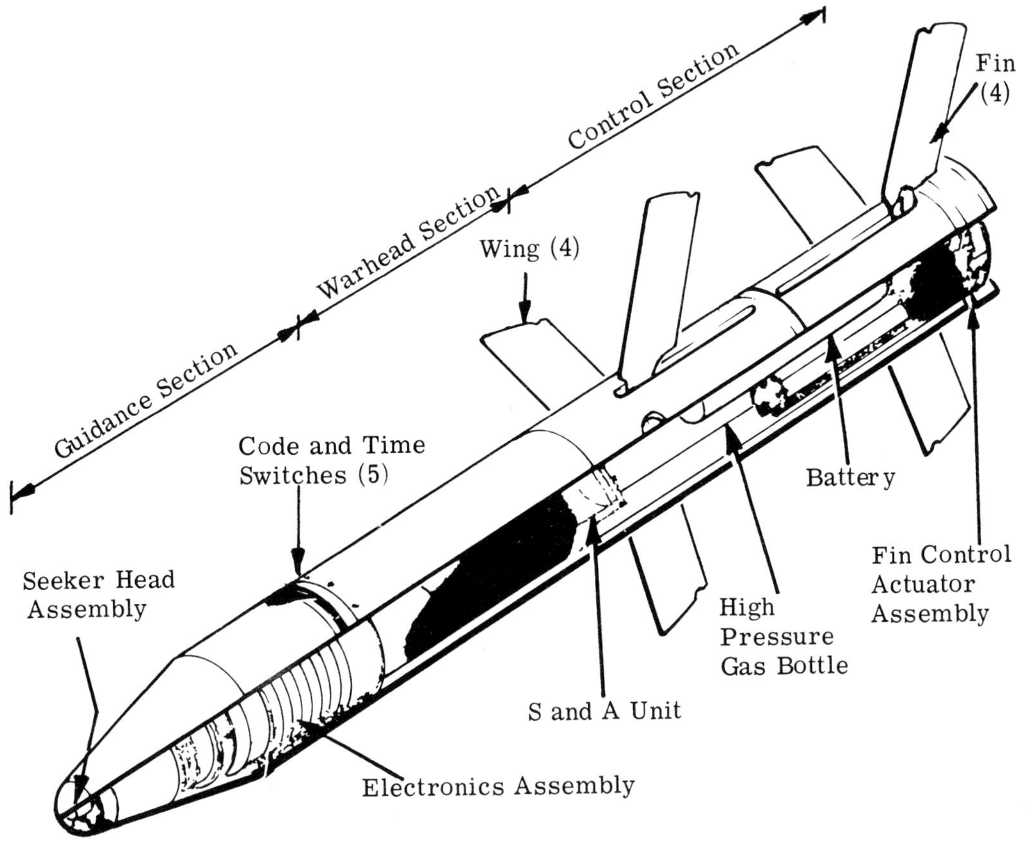

Fig. 7. A terminally guided munition

This type of munition requires a 'designator' to indicate and fix the target for the projectile to home in on. Copperhead uses a laser guidance system which looks for the laser reflections from the target which is being designated, as shown in Fig. 8. A full description of the round and its problems is given in Volume II in this series of Battlefield Weapons Systems and Technology books. Although

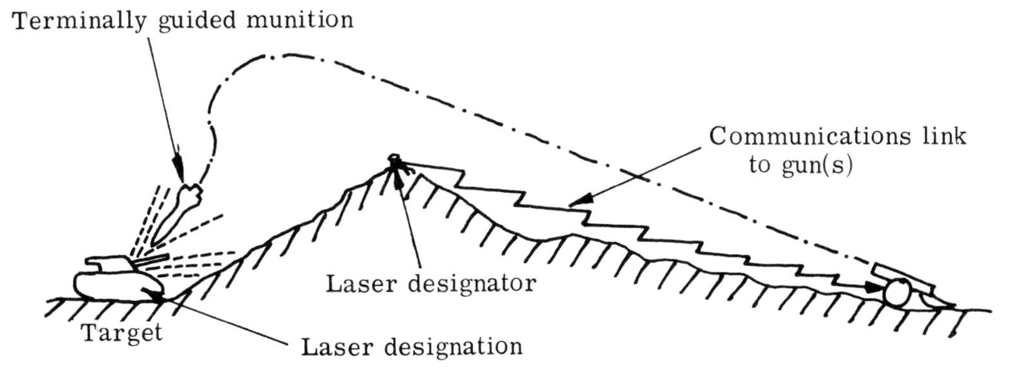

Fig. 8. Laser designation for a terminally guided munition

Copperhead employs laser seeking, other methods being considered include infra-red or magnetic passive homing systems, but problems associated with target discrimination have yet to be solved. TGMs, clearly, are complex and expensive, and it must be remembered that they depend on additional and expensive target illuminating systems which require a dedicated communications link with the guns firing the TGMs if the system is to be really effective. A TGM projectile must incorporate not only a warhead (usually HEAT) but guidance and control electronics, a gyro, power supplies, and an aerodynamic or impulse control system. The projectile must also be capable of withstanding the firing stresses in the gun. The performances of first generation TGMs are degraded by natural or induced conditions such as bad weather, enemy weapons fire, smoke and dust and they are fairly easy to counter by jamming and other electro-optical measures aimed specifically at defeating the guidance link. Nonetheless, despite the limitations of first generation TGMs like Copperhead, there is considerable potential for further development in this direction.

SUMMARY

There have been significant increases in the lethality, range and accuracy of conventional artillery fired ammunition in recent years. The introduction of improved conventional munitions with their enhanced performance against hard and semi-hard targets, means that indirect fire artillery can now engage armoured targets more effectively. Extended range projectiles give conventional artillery the ability to engage targets out to ranges of 30 kms or more, thus increasing its flexibility and survivability. The development of terminally guided munitions indicates the potential for a further enhancement in the effectiveness of indirect fire projectiles.

SELF TEST QUESTIONS

QUESTION 1 Define "Improved Conventional Munitions".

Answer ....

....

....

QUESTION 2 What is the main dilemma to be resolved in achieving enhanced effectiveness with sub munition carrying projectiles?

Answer ....

....

....

....

QUESTION 3 What are the options available for extending the range of projectiles?

Answer ....

....

....

....

QUESTION 4 Explain the term "pressure break point".

Answer ....

....

....

....

QUESTION 5 Describe, with the aid of a diagram, an "aerodynamically optimised" projectile.

Answer

....

....

....

....

....

QUESTION 6 What is "base bleed"?

Answer ....

....

....

QUESTION 7 Describe the method of reducing drag by "external bleeding".

Answer ....

....

....

....

QUESTION 8 Describe how the rocket assisted projectile works.

Answer ....

....

....

QUESTION 9 What are the advantages and disadvantages of ERPs?

Answer ....

....

....

....

QUESTION 10 What are PGMs / TGMs, and why do we need them?

Answer ....

....

....

....

ANSWERS ON PAGE 263

Answers to Self Test Questions

Page 13

QUESTION 1 Ammunition can be defined as any munition of war whether defensive or offensive, or any component whether filled or intended to be filled with explosive, smoke, chemical, incendiary, pyrotechnic or any other substance designed to affect the target, including any inert or otherwise innocuous training, practice or drill replicas and expedients.

QUESTION 2 The most effective conventional means of disabling a target is to apply energy to it in some way and as rapidly as possible.

QUESTION 3 An explosive is a substance with considerable potential energy relative to its mass which can be released when suitably initiated.

QUESTION 4 "High Explosives" used to achieve target effects; and "Low Explosives" used, more generally, as propellant charges.

QUESTION 5 An explosion is a violent expansion, usually of gaseous matter, and the energy of expansion appears primarily in the form of heat, light, sound and a shock wave.

QUESTION 6 A "High Explosive" is one which "detonates" rather than "burns" (Low Explosive).

QUESTION 7 Igniferous (Burning) train.
Disruptive (Detonating) train.

QUESTION 8 Brisance is the "shattering" property shown by an explosive. It is the product of the power and velocity of detonation of the explosive concerned.

QUESTION 9 Obturation is the prevention of the uncontrolled escape of propellant gases. It is the sealing of the breech of a weapon to ensure that all the propellant gases act on the base of the projectile to send it on its way to the target, and that none of the gases can escape to the rear until the projectile has left the barrel and the breech is opened. Obturation also involves preventing the forward escape of the gases past the projectile and this is one of the functions performed by the driving band.

QUESTION 10 From the middle of the 15th century until the middle of the 19th century the great guns and cannons (artillery) were controlled by the Board of Ordnance. Hence the guns themselves became to be referred to as pieces of "ordnance".

CHAPTER 2

Page 23

QUESTION 1 $P_K = P_H \times P_R \times P_L$ (Where P_K = chance of a kill; P_H = chance of a hit; P_R = reliability of the complete weapon system; P_L = lethality of the warhead). Such an expression is necessary as it provides a framework against which the effectiveness of a complete weapon system can be measured and compared with other systems.

QUESTION 2 Neutralisation.
Disablement.
Destruction.

QUESTION 3 Disablement of the target to prevent it from doing its job is the damage level the ammunition designer aims to achieve. Neutralisation applies to people; material cannot be neutralised, and destruction is seldom cost effective since it leads to substantial overkill on many targets.

QUESTION 4 People.
Armoured vehicles.
Structures.
Equipment.
Aircraft in flight.

QUESTION 5 Yes, but they are normally discounted. Psychological and physiological effects are of a transient nature and are unlikely to achieve any significant level of disablement against well led and disciplined troops. Hence, it is normal practice to discount such effects when assessing ammunition effectiveness.

QUESTION 6 The ammunition.
The means of projection and delivery.
Development and production.

QUESTION 7 Kinetic Energy (KE)
Chemical Energy (CE)

QUESTION 8 Man is delicate and vulnerable and provided he can be hit, he is comparatively easy to disable. Thus kinetic energy projectiles are effective, but so too are blast and fragment (CE) projectiles.

QUESTION 9 Anti-armour ammunition must first be capable of defeating the armoured vehicle's protection and still retain sufficient residual energy to do damage behind the armour to the crew, equipment, and machinery.

QUESTION 10 If kinetic energy is used, the aircraft must be hit if the target effect is to be achieved. Alternatively, a combined kinetic energy / chemical energy warhead, to produce blast and fragments, can be effective against an aircraft without actually hitting it.

CHAPTER 3

Page 37

QUESTION 1 Small, about 0.42 m^2. Protected by clothing, flak jacket etc. In the open or behind cover. Fit or unfit, alert or tired, well motivated or depressed.

QUESTION 2 Standing, palms forward, facing the attack. Frame identified through 108 slices taken horizontally. Wound tract being applied in each of six 60° paths.

QUESTION 3 High chance of a hit, high $mv^3/2$, rapid energy transfer, no unnecessary overkill.

QUESTION 4 Mass, velocity and shape.

QUESTION 5 Terminal effects and efficiency of a device.

QUESTION 6 Incapacitation within 30 seconds of the strike on a determined, well protected target in a good defensive position.

QUESTION 7 Initiated by a fuze, detonation wave expands and pressure fractures the device and propels fragments outwards.

QUESTION 8 The nearer to the perpendicular the more circular is the fragment zone.

QUESTION 9 An air burst enables most of the projectile fragments to be hurled at the target but a ground burst results in some fragments being lost to the ground.

QUESTION 10 Not so important or significant but it does contribute to casualty numbers. Sound can leave lasting effects.

CHAPTER 4

Page 53

QUESTION 1 A mechanical mixture of potassium nitrate / charcoal / sulphur 75/15/10.

QUESTION 2 High explosives detonate, propellants burn or deflagrate.

QUESTION 3 A system of building up various explosives by sensitivity to ensure efficient functioning with safety.

QUESTION 4 The explosive train starter.

QUESTION 5 Igniferous or disruptive.

QUESTION 6 Limiting shock velocity of a particular explosive. Always exceeds the velocity of sound in the explosive.

QUESTION 7 It is a measure of an explosive's reaction to certain stimuli eg impact and friction. Compared with the standard Picric Acid which has a value of 100.

QUESTION 8 Single base - nitro cellulose.
Double base - nitro cellulose and nitro glycerine.
Triple base - nitro cellulose, nitro glycerine and nitro-guanadine, or picrite.

QUESTION 9 Various chemicals added to the basic ingredients to provide stability, plasticity, lubrication, or to give less flash, smoke or cooler burning. Examples are carbonite, potassium cryolite etc.

QUESTION 10 Where rocket propellants are made to provide stable and uniform burning to give a characteristic plateau in the pressure time curve.

CHAPTER 5

Page 69

QUESTION 1 Part of a round of ammunition which provides the means of propelling the projectile to the target.

QUESTION 2 Quick firing, breech loading.

QUESTION 3 The identification of the obturation system in a weapon currently denoted by a cased charge for QF and bagged charged for BL.

QUESTION 4 Forward momentum of the projectile is balanced by the rearward momentum of the propellant gases which are allowed to escape rearwards at a certain time.

QUESTION 5 The start of burning in a deflagrating chain.

QUESTION 6 Propellant contained in a bag which is totally consumed when fired. Propellant contained in a metal case, usually brass, which is removed after firing.

QUESTION 7 By an obturating pad in the weapon for a bagged charge system and by the cartridge case and primer in the cased charge type.

QUESTION 8 A tube is required to produce a flash and/or hot particles which travel through a vent to the bagged charge igniter. A primer produces a flash to the surrounding propellant via the magazine in a cased charge system.

QUESTION 9 Protect contents, provide obturation, allow easy loading and extraction, houses the primer, supports the projectile.

QUESTION 10 Use of cool propellants, various additives, high density polyurethane liners, fine talc and wax wrapped in dacron cloth, combustible cases, and titanium dioxide plus talc.

CHAPTER 6

Page 80

QUESTION 1 Cheap, easy and safe to manufacture, strong to stand up to firing stresses, stable and accurate in flight, provide efficient functioning at the target.

QUESTION 2 a. Set back, spin, driving band pressure, buffetting (side slap)
b. Stability, shape, carrying power related to air resistance and drag.
c. Impact, produce required fragment effects.

QUESTION 3 Centre the projectile in the bore, prevent forward escape of gas (obturation), prevent slip back of projectile when loaded at high elevations, rotation of projectile.

QUESTION 4 The weight of the explosive filling as a ratio of the total filled weight of the projectile.

QUESTION 5 A specification of all components used in a projectile and how they are assembled or filled.

QUESTION 6 To avoid contamination, ingress of moisture or deterioration of the filling.

QUESTION 7 Pouring and pressing.

QUESTION 8 Loose or cracked, porosity, piping, cavitation.

QUESTION 9 To ensure that, in the event of piping in the projectile base, the hot propellant gases cannot penetrate into the high explosive filling during firing.

QUESTION 10 Billet cut from steel bar; heating and piercing; drawing; centring; rough machining and heading; fuze hole boring; driving band grooving; fuze hole threads cut; waved ribs in driving band groove.

CHAPTER 7

Page 92

QUESTION 1 Efficiency, durability, lethality.

QUESTION 2 Internal blast type is designed to hit or penetrate the target before detonating. External blast type is designed to produce damage when detonated near the target.

QUESTION 3 Rarified air at altitudes reduces blast pressure value.

QUESTION 4 Natural fragments are produced by a metal case being expanded and broken naturally by high explosive. Pre-formed fragments are built up into a warhead shape, held together in resin or a thin case and are thus ejected in their original shape when appropriate.

QUESTION 5 To ensure correct formation of the hoop without fracture.

QUESTION 6 The importance of ensuring that the warhead effects are concentrated towards the target rather than in a 360° mode cannot be overstressed.

QUESTION 7 A bomblet usually carries its own initiation system and is distributed randomly when released from the parent warhead. A sub-projectile normally has its own means of propulsion which is used on ejection from the parent warhead.

QUESTION 8 Sensitivity, power, velocity of detonation.

QUESTION 9 A combination of mechanical and explosive devices linking the fuze to the warhead.

QUESTION 10 Forces include set back at launch, creep forward when flight is steady, little or no spin and all forces of a much lower order than traditional gun firing forces. Safety and arming mechanisms are designed accordingly and also often use stored energy devices and generated energy producers.

CHAPTER 8

Page 122

QUESTION 1

M (Mobility) Kill = AFV immobilised and incapable of executing controlled movement, and is irreparable by the crew on the battlefield.

F (Firepower) Kill = Main armament is put out of action, either because the crew of the AFV have been rendered incapable of operating it, or because the associated equipment has been damaged making the gun inoperative and irreparable by the crew on the battlefield.

K Kill = AFV destroyed, knocked out, immobilised and damaged beyond repair.

P (Payload) Kill = Applys to APCs/ MICVs only and refers to the percentage of the payload that are incapacitated and incapable of fighting on effectively.

QUESTION 2

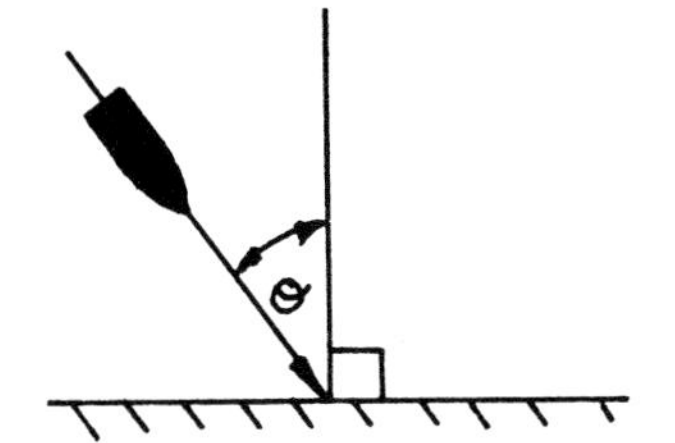

θ = angle of attack (ie the angle between the normal and the line of shot arrival).

QUESTION 3

The tank is a formidable target because it is fitted with armour plate of varying thicknesses and plate material, sloped at various angles with a variation of striking surfaces which include a variety of discontinuities such as spare track links, tools, lifting plugs, stowage bins and the such. All help to defeat the ammunition by absorbing and attenuating the energy or aiding ricochet. It also moves, rarely exposes itself completely and usually retaliates promptly.

QUESTION 4

Kinetic energy is a sheer "brute force" method of attack which involves projecting a solid shot at a target as hard and as fast as it is possible to do so.

QUESTION 5

$\frac{T}{d} = \frac{mv^2}{d^3}$ (Where T = thickness of plate
d = shot diameter
m = Mass
v = Velocity)

QUESTION 6

Swivel nose cap gives AP shot a better performance against sloped armour than would appear from the Cosine Law. The swivel nose cap starts to ricochet on impact, and in doing so, turns the nose of the shot towards the normal angle before it starts to penetrate (see Chapter 8, Fig. 4).

QUESTION 7

a. At the normal, the shot passes straight through the plate, causing front petalling, plugging and back petalling or a combination of these last two.

b. At high angles of attack the shot follows an S-shaped path through the plate; it wriggles its way through.

QUESTION 8

Barrelling:- Compression failure in soft steel; shot increases in diameter ("bellies") at its mid point and fails to penetrate.

Shatter:- To offset barrelling, the compressive strength of the shot is increased making it more brittle. On impact tensile hoop stresses cause the shot to break up or "shatter".

Lateral Bending:- At high angles of attack, shot is subjected to severe lateral stresses, both shear and bending, particularly towards its rear.

QUESTION 9

a. High value of $\frac{mv^2}{d^2}$; (a long thin projectile)

b. High value of $\frac{mv^2}{d^2}$; (a long thin projectile)

c. High value of $\frac{d^2}{m}$; (a short, squat projectile)

QUESTION 10

AP = armour piercing; C = cap to overcome shatter; BC = ballistic cap to improve shape to maintain velocity.

QUESTION 11

APFSDS: Advantages: High mass over small diameter = good penetration; can be used from rifled or smooth bore guns.
Disadvantages: Increased gun wear: marginally less accurate than APDS because it loses velocity down range more quickly and is more susceptible to cross wind effects.

QUESTION 12

The hollow charge effect is achieved by using the energy available from the detonation of a charge of high explosive to collapse and break up a metal liner into a metallic jet and plug. The velocity of the jet is about 9000 metres per second at its tip, and about 1000 metres per second at its tail, and it penetrates armour plate by the intense concentration of kinetic energy which is thus available at its tip (200 tons per square inch).

QUESTION 13

Good penetration = limited lethality (exit hole small).
Good lethality = large exit hole (but depth of penetration poor).

QUESTION 14

a. Cone diameter: less than 70 mm = poor lethality; more than 130 mm = good lethality; between 70 mm and 130 mm = a significant increase in lethality for every mm increase in cone diameter.

b. Spin: degrades the performance by diffusing the jet thus attenuating its concentration of kinetic energy at its tip.

c. Stand-off distance: the jet must have distance to stretch and concentrate to achieve optimum penetration.

d. Cone angle: normally between 40° and 80° and is usually 60°.
: large cone angle = poor penetration but good lethality (if penetration is achieved).
: small cone angle = good penetration but poor lethality.

QUESTION 15 HE is "pancaked" on to armour plate and detonated, sending a high velocity compressive shock wave through it. On reaching far side of plate, the shock wave is reflected back through the plate as a tension wave. When this rebounding tension wave meets further primary shock waves, they combine setting up a reinforced shock wave which exceeds the strength of the plate and a large scab is detached from the rear surface.

QUESTION 16 It is defeated by spaced armour.

QUESTION 17 Various forms of "plate charge" which are a cross between KE and CE. These are: the P (Plate) charge itself, the Misznay-Schardin plate and the implosive or "self forging fragment" plate charge.

Armour Piercing High Explosive (APHE) is yet another alternative form of attacking armour.

QUESTION 18 The essential difference between the various plate charges and HEAT is that as the angle at the "apex" of the liner (plate) becomes more obtuse, so the velocity of the leading part of the metallic slug that is formed decreases, and the velocity of the tail of the slug increases. Thus a slug only is formed, rather than a metallic jet and a plug, as is the case with HEAT.

QUESTION 19
: Poor penetration.
: Good lethality.
: No stand-off distance required.

QUESTION 20 APHE is essentially a kinetic energy solid shot, but with a base cavity filled with HE. As the shot penetrates the target, the HE filling is detonated and this enhances the effect behind the plate.

CHAPTER 9

Page 135

QUESTION 1 F_t, C_t, E_t

QUESTION 2 Each element (vulnerable area) is assessed separately for each damage criterion and by relating each to the total presented area of the target, the chance of a kill can be found.

QUESTION 3 Armour protection, burying, duplication.

QUESTION 4 Gun fired projectiles cannot change direction after firing, missiles, therefore, are more able to cope with fast targets.

QUESTION 5 Blast, fragments, shaped charge.

QUESTION 6 Larger cone angle and cone diameter, aluminium cone of thicker profile, long and fat jet.

QUESTION 7 To produce incendiary effect and to cope with miss distances.

QUESTION 8 For cost effectiveness a warhead is needed for miss distance requirements although it may not be required if the missile is designed to hit the target.

QUESTION 9 There is little point in producing a system which operates on a miss distance of say 1 metre when the warhead has an effective lethal radius of 5-10 metres.

QUESTION 10 The air is rarified and cannot transmit pressure as well as dense air found at ground level.

CHAPTER 10

Page 150

QUESTION 1 Bursting; base ejection; nose ejection; base emission.

QUESTION 2 White phosphorus ignites spontaneously when projectile is broken open. Red phosphorus is ignited by the high explosive column or surround.

QUESTION 3 A time fuze initiates a burster charge which provides pressure to eject the contents through the base of the projectile. At the same time the contents may be initiated as appropriate.

QUESTION 4 Central or end ignition.

QUESTION 5

	Bursting	BE
Smoke produced quickly	Yes	No
Duration good	No	Yes
Pillaring	Yes	No
Effects in soft ground	Buries	Buries in soft snow
Ranges with HE projectile	Yes	No
Smoke produced where projectile lands	Yes	No
Secondary effects	Yes	No
Economical production	No	Yes
Safety	"Leakers"	Yes
Independent of atmospheric humidity	Yes	No

QUESTION 6 By two steel semi-cylinders.

QUESTION 7 Shrapnel, canister, flechette.

QUESTION 8 Sphere, cube, near cube, glider and wedge.

QUESTION 9 High explosive substitute used in practice projectiles.

QUESTION 10 A thin metal cased projectile containing a number of similar steel fragments either spherical or cylindrical which are propelled from the gun muzzle in a dense conical pattern.

CHAPTER 11

Page 166

QUESTION 1 A device used to initiate an explosive store.

QUESTION 2 Safety, reliability, durability.

QUESTION 3 By position, function or filling.

QUESTION 4 Acceleration, deceleration, spin, side slap.

QUESTION 5 Holding, masking and firing devices.

QUESTION 6 Safety throughout the logistic system, during firing and up to the time when arming is required.

QUESTION 7 Yes. Explosive components such as detonators, pellets, stemming and magazines.

QUESTION 8 A device for operating the fuze after a pre-set time either mechanical or combustion, and a mechanism which operates on impact with the target.

QUESTION 9 An automatic fuze which operates when a transmitted signal is received back from the target at a predetermined intensity.

QUESTION 10 A normal pressure type fuze which operates when the pressure of a tank is applied to it.

CHAPTER 12

Page 178

QUESTION 1 In a mortar the recoil force is usually transmitted directly to the ground through the base plate. In a gun it is transmitted through a recoil system.

QUESTION 2 By fins.

QUESTION 3 To provide a method of charge adjustment for different range settings. Cannot contain all propellant in a primary cartridge except for small calibres.

QUESTION 4 By expanding plastic obturator rings.

QUESTION 5 HE, smoke, illuminating, practice, drill.

QUESTION 6 By producing a strong one-piece tail unit.

QUESTION 7 Safety pin, long travel detent.

QUESTION 8 By muzzle loading, speedy clearance of barrel after each round is fired, smooth bore and expandable obturator ring.

QUESTION 9 It behaves more like steel producing better lethal fragments.

QUESTION 10 Because the modern mortar is accurate and similar in terminal effects to a close support field artillery gun.

CHAPTER 13

Page 191

QUESTION 1 A typical round consists of a cartridge case containing a cap and propellant plus a bullet attached at the front end of the case.

QUESTION 2 To reduce the weight thus allowing more rounds to be carried. Increased muzzle velocities have also allowed smaller bullets to be used.

QUESTION 3 To protect the lead core and prevent it fouling the bore.

QUESTION 4 Velocity required, weight and bulk, strength, safety and reliability.

QUESTION 5 Solid bullets are used for anti-personnel and armour piercing roles. Filled bullets comprise mainly tracer or incendiary compositions.

QUESTION 6 To hold the bullet and attain correct bullet pull value, provide obturation, contain and protect the propellant and initiation system.

QUESTION 7 The method of initiation generally involves a striker in the bolt of the weapon impinging on a small cap filled with sensitive explosive mixture. This action squeezes or traps the composition between the cap and anvil thus causing initiation. The flash produced ignites the propelling charge.

QUESTION 8 Integral anvil is part of the cartridge, separate anvil is assembled with the cap both types providing the squeezing effect on the explosive composition. The rimfire system used mainly in .22 inch has no cap or anvil but utilises the rim of the cartridge case.

QUESTION 9 Location of the initiating cap and extraction, cook-off, consistancy in pressure/time performance.

QUESTION 10 Due to the requirements for a modern soldier to carry additional items into battle, vizor, flak jacket, etc, it is essential to reduce the weight of ammunition carried which in turn could reduce the weight of the weapon also.

CHAPTER 14

Page 202

QUESTION 1 It comprises a container filled with high explosive, plus a fuze system incorporating a safety device and a delay. The container provides the fragments or may contain fragments which are projected when the grenade is initiated by a detonator.

QUESTION 2 High explosive, chemical, pyrotechnic.

QUESTION 3 By the use of a safety pin which is removed by the user when throwing or firing.

QUESTION 4 Projectors, discharger cup, spigot attachment and bullet trap utilised the normal small arms round or a special cartridge in a rifle.

QUESTION 5 Offensive grenades are used where the thrower is not protected and utilises the effects of noise and blast. Defensive grenades have larger lethal areas and require the thrower to be behind cover.

QUESTION 6 Safety in handling and carrying, reliability, waterproof, consistant.

QUESTION 7 The defeat of modern armour requires a larger grenade than is possible to be carried on the person.

QUESTION 8 Carrier grenades contain various fillings other than high explosives, such as smoke, gas, flare and incendiary. These grenades are designed to screen targets, provide anti-riot irritants, give signals, etc.

QUESTION 9 A ring aerofoil grenade has greater accuracy and range due to the lift provided by its hollow doughnut shape.

QUESTION 10 A grenade which can be used in the defensive or offensive role, filled with high explosive or smoke composition (for example) is known as a multi purpose grenade. Greater flexibility can be obtained if it can be thrown or projected.

CHAPTER 15

Page 221

QUESTION 1 A case, a fuze, an explosive train, and an explosive filling.

QUESTION 2 Lethality, sensitivity, safety, ease of laying, reliability, resistance to countermeasures, camouflage, logistically economic, and cheap. Additional characteristics which may be required are delayed arming, self sterilisation, remote control, and remote sterilisation.

QUESTION 3
: Tracks.
: "Belly" (underpart of a tank).

QUESTION 4
: Single impulse fuze.
: Double impulse fuze.
: Anti-disturbance fuze.

QUESTION 5
: Individual attackers.
: Mass attackers.

QUESTION 6 Anti-personnel mines can be hazardous to lay, and are difficult and dangerous to clear. They are also indiscriminate and a hazard to own forces, the civilian population and animals and will remain so for many years unless a reliable self-destruct or self-sterilising element is contained in the fuze.

QUESTION 7
: Hand laying.
: Mechanical laying.
: Surface delivery from a gun, rocket, mortar or dispenser tube.

QUESTION 8 The anti-tank mine plough skims off the earth in front of the tank tracks and rolls the earth away to one side with the mines that it contains. The fitting of a plough to a standard gun tank can impose considerable strain on the steering and transmission clutches of the tank. Ploughs can be defeated by deep buried mines, and mines with anti-disturbance fuzes (although these merely destroy the plough rather than the tank).

QUESTION 9 Propelling, by rocket, an explosive hose across the minefield and detonating it. The blast activates or physically disrupts the mines and clears a path through the minefield.

Minefields can also be disrupted by the use of fuel/air explosives, contained in shells, mortar bombs, or sprayed by aircraft, to produce blast overpressures of sufficient magnitude to detonate buried or surface laid mines.

QUESTION 10
: Tactical minefields. Sited to deflect enemy into prepared killing/ambush areas.

: Protective minefields. Small minefields laid by units from their own resources for local protection.

: Phoney minefields. Marked and fenced as minefields, but do not contain mines.

: Nuisance minefields. Small minefields designed to delay and disorganise enemy forces, by mining in depth along their likely main axis during the withdrawal of our own forces.

: Minimal minefields. Hastily laid minefields of a few rows only, possibly surface laid because of the lack of resources or time to lay tactical minefields.

CHAPTER 16

Page 231

QUESTION 1 Illuminants, smoke, incendiary, signal, tracer, delay and priming compositions.

QUESTION 2 An intimate mixture of fuel and oxidant plus additives as appropriate.

QUESTION 3 Binders improve cohesion of powdered ingredients, protect metal powders from corrosion and reduce sensitivity. Waxes, resins, oils and varnish.

QUESTION 4 A photoflash is a short high intensity effect for photography. An illuminating flare provides sustained light for illuminating the target or battlefield.

QUESTION 5 These are limited generally to the distinct colours which stand out in battle conditions and are needed to provide flexibility in signaling, indicating and so on.

QUESTION 6 Bursting type gives instant smoke plus incendiary effect but tends to pillar due to heat of combustion. Screening type is slower to build up but gives more sustained smoke, it is affected by atmospheric conditions but is cheaper.

QUESTION 7 Metals which burn in air such as aluminium or magnesium and combustible fluids which spread over the target while burning such as petrol or napalm.

QUESTION 8 They are used to show the trajectory or line of flight of a projectile.

QUESTION 9 Gas producing delays usually contain organic matter and produce large volumes of gas. Gasless delays produce mainly solid residue and can be used in confined spaces.

QUESTION 10 Pyrotechnic devices to portray actual events. Used for simulating weapon fire and effects or as training aids for sound, vision etc.

CHAPTER 17

Page 243

QUESTION 1 Improved conventional munitions are designed to give an improved performance compared with standard HE shell against hard and semi-hard targets.

QUESTION 2 The dilemma to be resolved in achieving enhanced effectiveness compared to HE is between delivering sufficient sub-munitions to give a satisfactory chance of a hit, and at the same time having devices that are large enough to do sufficient damage when they hit.

QUESTION 3
: Improve weapon performance.
: Improve ballistic coefficient.
: Provide some form of post firing boost.

QUESTION 4 "Pressure break point" is the point at which a further increase in propellant charge does not impart more energy to the projectile. The increased energy generated is expended in accelerating the increased mass of gas.

QUESTION 5

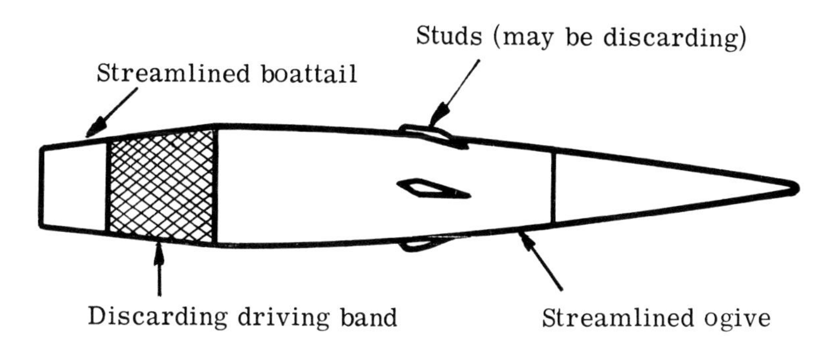

QUESTION 6 "Base bleed" involves burning a quantity of propellant, which is fitted into the rear of the projectile, at low pressure so that it generates sufficient gas to fill the partial vacuum that is always present at the base of a projectile while it is in flight. By filling this vacuum, base drag is reduced by up to 50%, giving the projectile a range increase of between 10% and 20%.

QUESTION 7 "External bleeding" involves burning a quantity of propellant set into the base of a projectile, with the gases produced being injected transversely through holes around the periphery of the base into the boundary layer of recirculating, turbulent air flow at the rear of the projectile.

QUESTION 8 A solid fuel motor is incorporated in the rear part of the projectile, and is lit by the main propellant charge during the firing process. The base of the projectile contains a rocket nozzle through which the motor gases pass, giving a net thrust.

QUESTION 9 Extra range is achieved at the expense of accuracy, payload, increased cost and increased target acquisition costs.

QUESTION 10 Precision guided munitions/terminally guided munitions. PGMs/TGMs are required to make conventional artillery more effective by achieving more impressive kill probabilities. In other words by having the ability to guide a projectile onto a target, "every round can be made to count".

Glossary of Ammunition and Explosives Terms and Abbreviations

A

AA - Anti Aircraft.

ACM - Anti Armour Cluster Munition (USA).

Adaptor - A threaded bush used to adapt a fuze to a projectile or bomb.

AFV - Armoured Fighting Vehicle.

Amatol - Ammonium nitrate and TNT mixed in various proportions (eg 80/20) for use as a shell filling.

Ammo - Ammunition.

Ammunition - Any munition of war or any component whether filled or charged or intended to be filled or charged with explosive, smoke, chemical, incendiary or pyrotechnic compositions; also,

a. Explosives in made up charges.
b. Explosives, chemical chargings, and incendiary, smoke or pyrotechnic materials in bulk.
c. Non explosive projectiles of all natures.
d. Non explosive stores and components for use in the initiation or assembly of projectiles or explosive charges.
e. Improvised explosive devices.
f. Home made explosives.
g. Guided Missiles and explosive components designed to be fitted to guided missiles.

The term 'Ammunition' is often used for inert replicas of the above.

Angle of Attack (or angle of incidence) - The angle between the line of arrival and the normal to the surface of impact.

UK usage implies θ

Some continental usage implies α

Anvil - That part of a percussion initiation system on which the cap composition is compressed.

AP - Armour Piercing. Usually a solid shot.

APC - Armoured Personnel Carrier.
- Armour Piercing Cap.

APCBC - Armour Piercing Cap Ballistic Cap.

APCNR - Armour Piercing Composite Non Rigid.

APCR - Armour Piercing Composite Rigid.

A Pers - Anti-personnel.

APDS	-	Armour Piercing Discarding Sabot (Projectile).
APFSDS	-	Armour Piercing Fin Stabilised Discarding Sabot (Projectile)
APHE	-	Armour Piercing High Explosive (Projectile).
API	-	Armour Piercing Incendiary (Projectile).
APSE	-	Armour Piercing Secondary Effects. Armour Piercing projectile containing a small quantity of chemical (usually red phosphorus) which on impact with the target will increase the behind armour effects.
Arm	-	The preparation of a weapon or munition for functioning.
Arming	-	(Fuzes and Safety and Arming mechanisms). The removal of munition initiation safety devices (often in sequence).
AT	-	Anti-Tank.
ATGW	-	Anti-Tank Guided Weapon.
Attenuate	-	To reduce in force or value. The term is often used to describe the weakening of a Shock Wave by decreasing its amplitude or duration.
Aug Cart	-	Augmenting Cartridge. Additional charge increments in mortar ammunition, sometimes referred to as secondary charges.

B

Ballistic Cap	-	A cap fitted to a projectile to improve its ballistic properties.
Base	-	In conventional driving band projectiles, the part of the projectile in rear of the main driving band. In rear skirt banded projectiles, the portion of the projectile supporting the skirt band. In HESH projectiles the adapter which carries the base fuze.
Base Bleed	-	A method of generating gas, to fill the partial vacuum at the base of a projectile during flight, in order to reduce drag and thereby increase range.
BE	-	Base Ejection. A means of ejecting the payload of a carrier shell during flight, eg smoke cannisters, illuminating candle etc by the use of a small explosive charge.
Base Fuze	-	A fuze located in the base of a projectile.
Base Plate	-	A metal disc secured to the base of a high explosive shell to prevent hot propellant gases from coming into contact with the bursting charge in the event of piping.
Body	-	The empty projectile, with driving band, if applicable, before filling with the payload, tracer or the assembly of any other components.
Body Relief	-	That portion of the cylindrical part of the projectile body which is machined to a limited smaller diameter.
Bourrelet	-	A carefully machined portion of the projectile body between the ogive and the relieved section.
BL	-	The symbol for a system of rear obturation in which the sealing is achieved by the arrangement of the breech mechanism at the rear of a gun's chamber. Originally an abbreviation for 'Breech Loading'.
Black Powder	-	Gunpowder (USA).

Blank - Usually applied to a gun cartridge containing an explosive charge which gives a loud report when initiated; but not used to fire a projectile.

Blast - A destructive wave produced in the surrounding atmosphere by an explosion or detonation. The blast includes a shock front, high pressure gas behind the shock front and a rarefaction following the high pressure.

Blasting Cap - In the United States, this term refers to a demolition detonator.

Blasting Explosive - Explosive used for industrial mining, quarrying, etc.

Blind - Usually applied to a projectile, which due to a failure in the explosive train, fails to function after firing/throwing.

Boattailed - The tapered shape of the rear of a projectile which reduces turbulence and drag.

Bomb - An aerodynamically stabilised munition dropped from an aircraft, or fired from a mortar. Also gsed generally to cover types of improvised explosive devices (IED).

Booster - An explosive charge used to boost the small output from an explosive initiator. See 'Gaine' and 'Intermediary'.

BW - Bridge Wire. A thin resistive wire situated in heat sensitive explosive. Heating by the passage of an electric current will initiate the explosive.

Brisance - The shattering property shown by a detonating explosive.

Burning - The propagation of combustion (q . v.) by a surface process.

Burster - Small explosive charge within a munition designed to eject contents or to break open the munition.

C

Calibre - The nominal diameter of the bore of a gun measured across the lands of the barrel.

Canister Shot - A metal cylinder containing pre-formed fragments which are scattered as the casing disrupts on firing.

Cap - A small metal container filled with a flame-producing explosive composition.

Capacity - Sometimes known as charge weight ratio or CWR.

Cap Conducting - An explosive initiator which ignites when an electrical current is passed through it.

Cart - Cartridge. A cased quantity of explosive complete with its own means of ignition.

Carrier Shell - A shell which carries a payload, designed to eject the contents.

CC - Cap Conducting.

CE - Composition Explosive (Tetryl). An explosive usually used as an intermediary.
- Chemical energy.

Charge - A bagged, wrapped or cased quantity of explosives without its own integral means of ignition. Secondary means of ignition may or may not be incorporated.

Charge/Weight Ratio (CWR)	-	The ratio of the weight of explosive to the total weight of the munition.
CLGP	-	Cannon Launched Guided Projectile. A gun fired projectile incorporating a terminal guidance system (usually semi active homing).
Combustible Case	-	A gun propelling charge with an outer container made up of explosive/combustible material which is totally consumed on firing.
Combustion	-	An exothermic oxidation reaction producing flame, sparks or smoke.
Command Break Up Unit (CBU)	-	An explosive device in a guided missile which will break up the missile on command.
Conducting Compositions	-	A combination of a small quantity of graphite mixed with a sensitive initiatory explosive in order that the composition may be initiated electrically.
Continuous Rod	-	A guided missile warhead configuration, consisting of a series of steel rods longitudinally encasing an explosive charge. The rods are joined at alternate ends and create an expanding hoop on warhead detonation.
CS	-	A chemical anti-riot smoke composition.
CVT	-	An old term for a proximity fuze using a bracket arming sequence.
CW	-	Chemical warfare.

D

Defensive Grenade	-	A high explosive fragmentation grenade with a large lethal area, designed to be used from behind cover.
Deflagration	-	A rapid burning.
Detent	-	A mechanical, usually 'g' operated, locking device used in fuzes and safety and arming mechanisms.
Detonation	-	An exothermic reaction wave which follows and also maintains a shock front in an explosive.
Detonator	-	An explosive device for starting detonation. It is usually small and may be set off by impact, friction, electricity, flame, heat, etc.
Direct Action (DA)	-	A fuze which is designed to function instantaneously on impact. (USA - Super Quick).
Double Impulse (DI)	-	A mine fuze action, whereby the first stimulus (pressure) arms the fuze and the second initiates it.
Drill Ammunition	-	An inert replica of ammunition specifically manufactured for drill and instructional purposes.
Driving Band	-	A malleable or pre-engraved band pressed round a projectile which, when engaged in the rifling of the barrel, imparts spin to the projectile. It also prevents forward escape of the propellant gases, centres the projectile in the barrel, and provides shot start pressure.
DS	-	Discarding Sabot. A vehicle for a projectile which is designed to be discarded on leaving the bore of a weapon.

It provides support and forward obturation to a smaller calibre or unconventionally shaped projectile.

E

ECM - Electronic Counter Measures.
ECCM - Electronic Counter Counter Measures.
EDB - Extruded Double Base (Propellant).
EED - Electro' Explosive Device. A device which converts electrical energy into chemical energy by the application of suitable electrical stimuli.
EFC - Equivalent Full Charge.
EIED - Electrically Initiated Explosive Device.
EOD - Explosive Ordnance Disposal.
ERFB - Extended Range : Full Bore.
ER/RB - Enhanced Radiation/Reduced Blast Weapon. (Neutron Bomb).
Escapement - The part of a mechanism (in fuzes and safety and arming mechanisms) which controls the speed at which an arming mechanism operates.
Exploder - An intermediary high explosive charge designed to amplify the detonation wave produced by a fuze or detonator, to detonate the explosive filling of a shell or demolition charge. It may be in the form of a pressed pellet or filled container.
- A portable electric generator designed for setting off initiators.
Exploding Bridge Wire (EBW) - A bridge wire in a relatively insensitive explosive which, when a very high energy electrical impulse is passed through it will deliver enough energy to initiate the explosive.
Explosion - An exothermic reaction which takes the form of an extremely rapid combustion accompanied by the formation of large quantities of heat-expanded gas, producing sudden high pressure.
Explosive - A chemical substance or mixture which, when suitably initiated can react to produce an explosion.
Explosive Lens - A method of wave shaping or influencing the effect and direction of the shock wave from a detonation.
Explosive Power - The effective work done during the explosion or detonation of a standard quantity of an explosive in a prescribed test.
Explosive Train - a. The arrangement of explosive components used to lead an explosive reaction from one place to another.
b. A sequential arrangement of initiator, intermediate explosive (or booster), and main explosive filling.
External Ballistics - The science dealing with the motion of a projectile after it has been projected and before it strikes the target.
Exudation - The emission of any substance (oily, tarry, gaseous etc) from an explosive, generally due to thermal changes or chemical reaction.

EW - Electronic Warfare.

F

FAX - Fuel-Air Explosive.
Fixed Ammunition - Ammunition where cartridge case and projectile are fixed together. (Normally QF Fixed).
FFR - Free Flight Rocket.
FGA - Fighter Ground Attack Aircraft.
Flash Detonator - a. A small detonator which can be set off by flash or flame. It has no self-contained means of ignition. It is better referred to as a flash-sensitive detonator.
b. A small device with the appearance of a detonator but which produces flash rather than a detonation wave. This is more correctly called an igniferous initiator.
Flashless Propellant - A propellant which gives the minimum amount of muzzle flash.
Flechette - A small dart-shaped anti-personnel projectile.
F of I - Figure of Insensitivity.
Fragmentation - The shattering of a munition casing caused by an explosive charge.
FS - Fin Stabilised.
Fuse - A cord or tube containing an explosive train: usually slow burning.
Fuze - A device which is designed to ensure that ammunition is safe during storage, handling, deployment and projection, yet will initiate the explosive train at the required time when subjected to a suitable stimulus. In guided missiles: an electronic device which generates the firing signal for the warhead. It may contain electronic safety interlocks and need not be located next to the warhead. See also 'Safety and Arming Mechanism'.
Fuze, Electric - A form of electric igniter which is fired by electrical heating of a fine resistance wire (bridge-wire).
FWAM - Full Width Attack Mine. Anti-tank mine system.

G

g - gram.
G12 etc - Grain size of gunpowder compositions.
Gaine - An explosive train arranged to achieve, maintain or enhance detonation. Sometimes a fuze subassembly containing safety and arming devices, the detonator and the magazine.
Governing Section - The part of the shell forward of the driving band at which the stress on firing is a maximum.
GLGM - Ground Launched Guided Missile.
GP - Gunpowder.
GPMG - General Purpose Machine Gun.

Graze - A fuze action where the fuze mechanism is designed to function on significant retardation of the projectile. (A glancing blow will sometimes suffice).

Grenade - Ammunition containing a filling of explosive, smoke or other material, designed to be thrown by hand, projected by or from a small arms weapon or from a grenade discharger.

Guncotton - Nitrocellulose with a nitrogen content above 12.9%. Now obsolete as a demolition charge.

H

Hangfire - When the time interval between the initiation of a cartridge cap and the initiation of the propellant charge is excessive.

HE - High Explosive. Any explosive which is capable of detonation when suitably initiated.

HEAT - High Explosive Anti-Tank. Munitions, usually projectiles, using the shaped/hollow charge principle to penetrate a target. See: Monroe Effect.

HEP - High Explosive Plastic. USA equivalent to HESH.

HES - High Explosive Substitute. An inert filling used in some practice projectiles.

HESH - High Explosive Squash Head: a base fuzed anti-armour projectile.

Hexachlorethane (HCE) - The chemical base of many pyrotechnic screening smoke mixtures.

HF - High Frequency.

High Order Detonation - Detonation at a velocity approaching the maximum stable velocity of detonation for the system.

Hot Spot - A small, localised region in an explosive substance which is characterised by a temperature much higher than that of its surroundings.

I

ICM - Improved Conventional Munition (USA). An ambiguous term which usually refers to the "next generation" of munitions developed from those currently in use.

IED - Improvised Explosive Device. Non-service explosive munitions, usually contrived by guerilla or terrorist factions, designed to Kill, Destroy, Disfigure, Distract or Harass.

Igniter - A device for starting burning.

Ignition - The start of burning.

IHE - Insensitive High Explosive.

IHEP - Insensitive HE and Propellants (USA).

Inc - Increment (Gun and Mortar: Propelling charge). An additional propellant charge that can be added to the charges in a multi-charge system to achieve intermediate ranges.
An additional charge which can be added to an existing single charge for increased range/muzzle velocity.
(USA) The individual charges within a multi-charge system, either semi-fixed, or separate loading cartridges.

Inhibitor - a. A substance employed to prevent or oppose chemical change in a system.
b. An adherent layer of material applied to part of the surface of a propellant to prevent burning at that surface.

Initiator - A device for setting off explosives or pyrotechnics eg: a detonator or igniter. The term is also used as an abbreviation for the initiatory compositions used in these devices. Such compositions may also be known as primary explosives.

Instructional Ammunition - Inert replicas of ammunition (which may be sectioned) used for classroom instruction only.

Intermediary - A fairly sensitive high explosive used to boost the shock wave from the initiator and pass it on to the main filling. Such explosives tend to be classed now as main fillings.

Internal Ballistics - The study of the path of a projectile through the bore of a gun from 'shot start' until it leaves the muzzle.

Interrupter (USA) - A safety device in a fuze which prevents full arming until the projectile has left the muzzle of the gun.

IR - Infra-Red.

K

Kg - Kilogram.

KE - Kinetic Energy.

KTM - Kill Through Miss. A warhead not relying on a direct hit for its effect.

L

Lachrymatory - (Chemical smoke composition). A harassing agent causing copious weeping.

Lag Time - The period between first applying current to an electro-explosive device and subsequent operating of the device.

LAW - Light Anti-Tank Weapon.

LF - Low Frequency.

LMG - Light Machine Gun.

Long Rod Penetrator - A solid shot used for the attack of Armour. Usually made of steel, Tungsten or depleted Uranium alloys, with a length of up to 20 times its diameter; it is fin stabilised.

Long Travel Detent	- Detent found in mortar fuzes and others which travels a relatively long distance parallel to the longitudinal axis during firing.
LOVA	- Low Vulnerability Ammunition (USA): ammunition relatively insensitive to battlefield hazards such as fragment attack, fuel fires etc.
Low Explosive	- An explosive which does not detonate under normal conditions of use. This term is becoming obsolete.
Low Order Detonation	- Detonation of an explosive at a velocity well below the maximum stable velocity of detonation for the system.
LR	- Long Range.
LRATGW	- Long Range Anti-Tank Guided Weapon.

M

Magazine	- That portion of a fuze containing an intermediary explosive. - A building used to store ammunition. - The compartment of a ship where ammunition is stored. - An ammunition supply container fitted to automatic weapons.
MBT	- Main Battle Tank.
Mechanical Time Fuze	- A fuze incorporating a 'clockwork' mechanism capable of being pre-set to govern time of flight before functioning.
MICV	- Mechanised Infantry Combat Vehicle.
Mine	- A munition primarily designed to remain passive until it is initiated by a stimulus received from a target, time lapse or a 'command signal'.
Misfire	- The failure of an initiator or cartridge to fire.
Miznay-Schardin Effect	- Where detonating explosive is used to project a dished or coned plate at a target (usually armour).
ML	- Muzzle Loading.
MLRS	- Multi Launch Rocket System.
Mor	- Mortar.
MR	- Medium Range.
Multi Role Fuze	- A fuze with various operating options, ie Proximity, Direct Action, or Graze etc.
Munroe or Neumann Effect	- A concentration of the effect of a detonation by shaping the explosive charge; also called shaped charge. See 'HEAT'.

N

Necked	- The forward end of the cartridge case when reduced in diameter.
Neonite	- The proprietary name for ICI single base small arms propellant.
Nitroglycerine (NG)	- A sensitive viscous liquid explosive (Glycerol Trinitrate).

Nitrocellulose (NC) - Nitric esters of cellulose having a nitrogen content depending upon conditions of preparation; compare guncotton.

Nobel's (808, 851, 852 etc) - A series of British service plastic explosives of World War II vintage.

NPP - Nobel's Parabellum Powder. A porous, fast-burning variant of Neonite.

NRN - Nobel's Rifle Neonites; a numerous series of small arms propellants, each distinguished by a numeral following the initials.

O

Obturation - The sealing process which prevents propellant gases from escaping through the breech mechanism when a gun is fired. 'Forward obturation' prevents the propellant gases from escaping past a projectile until it clears the bore, usually by means of an obturating band on the projectile.

Offensive Grenade - A high explosive grenade of limited lethal area, designed for use by exposed troops. (See 'Defensive Grenade').

Ogive - That portion of a streamlined projectile starting at the front and continuing to the parallel or widest section.

OP - Observation Post.

P

'P' Charge - A plate charge variant of the Miznay Schardin effect.

Partial Ignition - Ignition in which the burning fails to propagate throughout the explosive.

PBX - Plastic-bonded explosive. Not to be confused with plastic explosive (PE): PBX's are mostly hard in their manufactured state.

Percussion Fuze - A fuze which functions on impact.

PETN - Pentaerythritol tetranitrate. (A high explosive).

PD - Point Detonating (Fuze).

PICRITE - Service code name for nitroguanidine, a major ingredient of triple-base propellants.

PGM - Precision Guided Munition (USA).

PGSM - Precision Guided Sub Munition (USA).

Plastic Explosives (PE) - The combination of an explosive with a plasticising material.

Platonising Agent - An additive incorporated into propellants in order to provide a constant burning rate.

PPD - Proximity Point Detonating Fuze.

Practice Ammunition - A munition designed to be used on training firings in lieu of the operational munition it replaces.

Premature - The premature functioning of a munition before its prescribed time of initiation; viz Bore premature, Flight premature.

Primary Explosive - A sensitive explosive which is readily initiated.

Primer - Intermediary explosive used to augment the impulse from the initiatory explosive to a magnitude sufficient to cause detonation of the main explosive charge. The terms exploder, gaine or booster are frequently used in connection with shells, bombs and torpedoes. In a gun cartridge, the explosive device containing a cap and a booster charge, which is used to ignite the main propellant charge.

Proj - Projectile. A missile projected by a gun or howitzer.

Proof - The assessment of performance of ammunition and explosive components by initiating and monitoring the results.

Prox - Proximity.

Propellant - Explosive used to propel a projectile or missile, or to do other work by the expansion of high pressure gas produced by burning.

Pyrotechnic - A chemical mixture of oxidising and reducing agents capable of reacting exothermically. Such mixtures are used to produce light, heat, smoke or gas and may also be used to introduce delays into explosive trains.

Pyrotechnics - A compound or mixture designed to produce an effect by heat, light, sound, gas or smoke or a combination of these as a result of a non-detonative self-sustaining exothermic chemical reaction.

Q

QF - The term for a system of rear obturation in which sealing is achieved by means of the cartridge case. QF 'fixed' ammunition: the cartridge case is firmly attached to the projectile. QF 'separate' ammunition: the cartridge case is separate from the projectile.

Quickmatch - An absorbent cord impregnated with a rapid burning pyrotechnic composition, usually gunpowder.

R

RAG - Ring Aerofoil Grenade.

RAP - Rocket Assisted Projectile. A projectile incorporating a rocket motor to increase its range.

Ramming Stop - With guns using separate loading ammunition, where more than one type of round is used, it is necessary, for ballistic regularity, to keep a consistent chamber capacity for one of the projectiles. This is done by machining a stop on the driving band to act as a ramming stop.

RCL - Recoilless. A projection system in which the forward momentum of the projectile is balanced by the rearward momentum of the escaping propellant gases.

Rd - Round. All the ammunition components necessary to fire a weapon once.

RD	- ('Research Department'). A series of materials, explosive or otherwise, which may be found in British service ammunition, each member of the series having a separate numerical suffix.
RE	- Radar Echo.
RDM	- Remotely Delivered Mine.
RDX	- Cyclotrimethylene trinitramine. (High Explosive).
RF Hazard	- (RADHAZ). The danger of the accidental initiation of an electro-explosive device by radio frequency electromagnetic radiation.
Rim Fire (RF)	- In small arms ammunition, a system of initiating the propellant charge by means of an annular charging of percussion sensitive mixture located in the inner cavity of the cartridge case rim.
Rotating Band	- USA, equivalent to driving band.
RP	- Red Phosphorus.

S

SAA	- Small Arms Ammunition, usually below 30 mm calibre.
Sabot	- A light weight full calibre casing for a small diameter sub projectile usually discarded at the muzzle.
SADARM	- Seek and Destroy Anti-Armour Munition (USA).
SAP	- Semi Armour Piercing (Projectile). A projectile or shot designed to penetrate light armour. Sometimes contains a high explosive filling.
Safety and Arming Mechanism (S & A)	- Used in guided missile warheads and weapon systems to describe a mechanical device containing the initiating explosive, a boosting charge and a mechanical shutter linked to safety interlocks. Its functions are: a. To provide warhead safety during transit, handling and firing. b. To provide a transition from 'Safe to Arm' at the appropriate time during flight. c. To initiate the warhead in response to a stimulus from the fuze.
Secondary Explosive	- A substance or mixture which will detonate when initiated by a shock wave or detonation wave, but which normally does not detonate when heated or ignited.
Self Destruct Unit	- A device which destroys a missile, either on command or automatically at an appropriate time.
Self Forging Fragment	- A projected Miznay-Schardin plate, usually a shallow dished plate, rather than a wide angled cone.
Sensitise	- To increase the sensitiveness or the sensitivity of an explosive.
Sensitiveness	- A measure of the relative ease with which an explosive may be ignited or initiated by a prescribed stimulus. (An inverse measure of safety of an explosive against accidental initiation).
Sensitivity	- A measurement of the relative ease with which reliable

		functioning may be assured in different explosives or explosive systems under the intended conditions of use.
Separated Case	-	A cartridge case which breaks circumferentially in the chamber of a gun, when the cartridge is extracted.
Set Back	-	The inertial forces acting upon the contents of a projectile due to its being accelerated on firing.
Set Up	-	The bulging of the rear part of the bullet due to the sudden pressure of the propellant gases and the inertia of the forward part of the bullet.
Shaped Charge	-	An explosive charge shaped to concentrate its effect when detonated (see Munroe effect). The effect of the charge is often enhanced by the addition of a metal liner.
Sheet Explosive	-	Plastic explosive in the form of a sheet.
Shell	-	A hollow projectile filled with a payload.
Shock Front	-	A discontinuous change in the pressure and other parameters of a medium. This change is propagated at supersonic speed.
Shock Wave	-	A shock front together with its associated phenomena (see Blast).
Shot	-	A solid projectile.
Shot Start Pressure	-	The pressure achieved by propellant gases at which a projectile starts to move.
Shutter	-	A safety device in an explosive train for isolating the initiating explosive.
SI	-	Single Impulse (Fuze).
Side Slap	-	The undesirable action of a projectile travelling through a bore of a gun where it moves from side to side and rebounds against the barrel.
Simulator	-	Ammunition device designed to simulate the firing of various weapons or exploding ammunition.
Smk	-	Smoke.
Smokeless Propellant	-	A propellant which leaves the minimum of residue of carbon deposit in the bore of a gun barrel, and gives little or no smoke on firing.
SP	-	Self Propelled.
Spun Tubular Projectile (STUP)	-	A projectile which is stabilised by spin and airflow through the aerofoiled tubular body.
Stabiliser	-	A substance which prevents or reduces auto-catalytic decomposition of explosives, especially propellants.
Stand-off	-	The distance of a shaped charge from a target at the instant of detonation. There is an optimum value at which best performance is achieved, usually expressed in charge diameters.
Stemming	-	An explosive filling which has been loaded in increments.
Striker	-	That part of a firing mechanism which hits an initiator.
Subcalibre Ammunition	-	a. A projectile which is of smaller calibre than that of the weapon as it leaves the muzzle. b. A practice projectile that is smaller in calibre than that which is usually fired from the gun.
Super Quick	-	USA equivalent to Direct Action. (Fuzes).

Sympathetic Detonation - The detonation of an explosive or of an explosive device as a consequence of another detonation. The term is normally used in accident reports.

T

T - Tracer. A pyrotechnic charge, fitted to the base of a projectile which, when ignited by propellant flash, burns with a bright flame and allows the firer to observe the flight of the projectile.

TGM - Terminally Guided Munition (USA).

TGSM - Terminally Guided Sub Munition (USA).

TNT - Trinitrotoluene. A High Explosive.

Triple Base Propellant - A class of gun propellants comprising nitrocellulose, nitroglycerine and nitroguanidine (picrite), the latter ingredient providing a reduction of gun-flash compared with other types of propellant.

Tube - A device used to initiate BL charges, usually fitted in the breech vent.

V

Velocity of Detonation (V of D) - This is the speed at which a detonation wave progresses through an explosive. When, in a given system, it attains such a value that it will continue without change, it is called the stable velocity of detonation for that system. (Commonly 6-10 km sec^{-1}).

VT - (Variable Time). A proximity fuze which is designed to function at a predetermined distance from the target. This term is normally superseded by "Proximity".

W

WAAM - Wide Area Anti-Armour Ammunition. (USA).

Wave Shaper - See Explosive Lens.

Windage - The difference between the diameter of the bore of a weapon and the diameter of a projectile.

Wind Shield (USA) - A cap over the head of a projectile to effect streamlining. (Equivalent to 'Ballistic Cap').

WP - White Phosphorus, mainly used as a rapid smoke producing agent.

WWI - World War I

WWII - World War II

Index